The Remarkable Life *of the* Skin

The Remarkable Life *of the* Skin

An Intimate Journey Across Our Largest Organ

Monty Lyman

Atlantic Monthly Press
New York

Text illustrations by Global Blended Learning
The Last Judgement by Michelangelo reproduced courtesy of
Granger Historical Picture Archive/Alamy stock photo.

First Published in Great Britain in 2019 by Bantam Press,
an imprint of Transworld Publishers

Published simultaneously in Canada
Printed in Canada

First Grove Atlantic hardcover edition: June 2020

Typeset in 11.75/15pt Caslon 540 LT Std by Jouve (UK), Milton Keynes.

Library of Congress Cataloging-in-Publication data is available for this title.

ISBN 978-0-8021-2940-6
eISBN 978-0-8021-4707-3

Atlantic Monthly Press
an imprint of Grove Atlantic
154 West 14th Street
New York, NY 10011

Distributed by Publishers Group West

groveatlantic.com

20 21 22 23 10 9 8 7 6 5 4 3 2 1

This book is dedicated to the millions across the world who suffer in, or for, their skin.

Contents

List of Illustrations

Author's Note: Discretion and Definitions

The Hippocratic Oath states, 'What I may see or hear in the course of treatment or even outside of treatment in regard to the life of patients, which on no account one must spread abroad, I will keep to myself.'[1] All doctors owe a duty of confidentiality to their patients, so every character with a skin condition in this book has been given a pseudonym. In some cases, particularly for those with very rare and identifiable skin diseases, I have applied a 'double lock' of anonymity, their name being changed and the location of the meeting moved, although always to a place I have either visited or worked in.

Although it is inappropriate to define someone by a disease, such as 'leper' or 'albino', I will occasionally use these terms to invite the reader into the daily reality of those suffering because of their skin.

Prologue

FOR A DOCTOR with an antiquarian bent, the magnificent Anatomical Theatre of the University of Bologna is heaven, even on an Italian summer's day, when the heat is unbearable and the wood-panelled hall effectively becomes a sauna. As I stood in the world's oldest university, whose four-hundred-year-old hall is carved entirely out of spruce, I felt like a shrunken man exploring the interior of an ornate and antique jewellery box. In the middle of the room lies an imposing marble dissection table, which for centuries provided medical students occupying the wooden stalls of this academic arena with a full view of proceedings. The walls are decorated with large, intricately carved wooden sculptures of the ancient heroes of medicine. Hippocrates and Galen examine the students below with forbidding glares – a look replicated by many subsequent medical school lecturers, no doubt. But, of all these wonders, the visitor's eye is drawn to the centrepiece. Overlooking the whole theatre is the seat of the professor, covered by an elaborate wooden canopy held aloft by the magnificent *Spellati*: the 'Skinned Ones'. These two statues, which take centre-stage at the altar to this church of medicine, are resplendent with exposed muscles, blood vessels and bones.

So-called écorché figures (from the French for 'skinned') depict the muscles and bones of the body and their interplay, but without skin. These sinewy, skinless bodies have been synonymous with medicine since Leonardo da Vinci's ground-breaking anatomical drawings of the fifteenth century, gracing the cover of almost every medical textbook. Peering up at the wooden écorchés

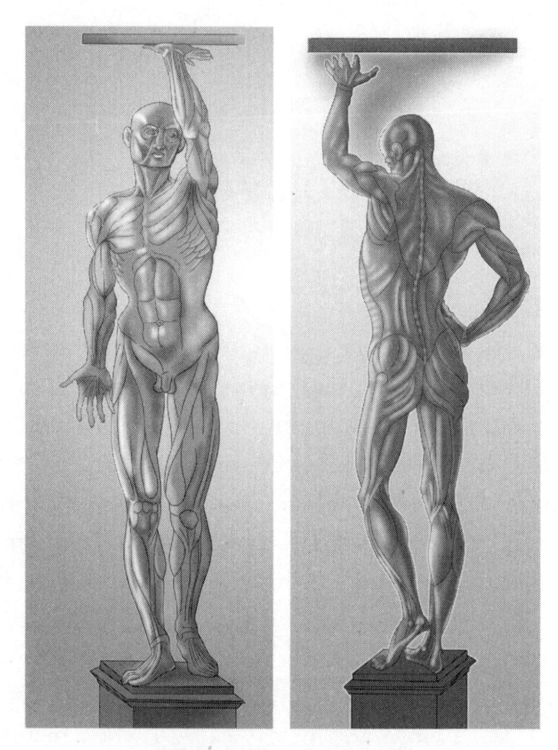

BOLOGNA'S *SPELLATI*

at the University of Bologna, it is clear to see that the skin – despite being our largest and most visible organ, despite us seeing and touching it, indeed living in it, every moment of our lives – is the organ most overlooked by the medical profession. Weighing nine kilograms and covering two square metres, skin wasn't even recognized as an organ until the eighteenth century. When we think of organs, or the human body at all, we rarely think of our skin. It is invisible in plain sight.

When new acquaintances ask about my clinical and research interests, my almost apologetic response, that dermatology excites me, is usually met with confusion, pity, or a combination of the two. A close friend who is a surgeon likes to taunt me by saying, 'The skin is the wrapping paper that covers the presents.' But part of what intrigues me is that, even though skin is the most observable part of our body, there is so much more to it than meets the eye.

My fascination with the skin started at the age of eighteen, on a slow afternoon two days after Christmas. My family had just made it through the last of the leftovers and I lay in a post-prandial sprawl on the sofa. Covered in blankets and revision notes, I sluggishly set about preparing for my first medical-school exam in a week's time. I felt a bit grotty and my inner elbows and face were feeling unusually itchy. A later look in the mirror revealed that my cheeks had gone a darker shade of pink. Within a couple of days, my face and neck had become a red, dry, itchy mess. My friends and family offered tellingly different explanations, from exam stress and house allergens, to over-hot showers, skin microbes and eating too much sugar. Whatever the reason, after eighteen years of clear skin, it suddenly started to break down – and eczema has shadowed me ever since.

Our skin is a beautiful mystery, cloaked in feelings, opinions and questions. The more science reveals of this terra incognita, the more we see that our most overlooked organ is actually our most fascinating. Skin is the Swiss Army knife of the organs, possessing a variety of functions unmatched by any other, from survival to social communication. Skin is both a barrier against the terrors of the outside world and – with millions of nerve endings to help us feel our way through life – a bridge into our very being. Simultaneously wall and window, our skin surrounds us physically, but it is also an exquisitely psychological and social part of our being. Our skin is not just a marvellous material; it is a lens through which we learn about the world and ourselves. Our physical skin teaches us to marvel at the intricacy of our bodies and the wonders of science. It teaches us to respect the millions of microbes that accompany us on our journey, to be sensible – not radical – about what we eat and drink, and to revere, but not fear, the sun. Our ageing skin directly confronts us with our own mortality. The mind-boggling sophistication of human touch invites us to re-examine the role of physical contact in an increasingly isolated and computerized society. There is no better platform than the psychological skin to show how our body and mind, and indeed our physical and mental health, are inextricably linked. Clothing, make-up, tattoos, society's heated dialogue on skin colour, and the judgement millions have suffered for skin deemed to be diseased or dirty, show that skin is our most social organ. Finally, ultimately, our skin transcends its physical presence and influences our faith, language and thinking.

The Remarkable Life of the Skin is not a step-by-step guide to beautiful or healthy skin. Although you will find information about how you can care for your outer surface, this book is perhaps more important than that. It is a circumnavigation of, and a love letter to, our most remarkable organ. Using the skin as a prism, what follows is a look across both time – from ancient history to the future of science – and space – from the elegant tattoos of crocodile-worshippers in Papua New Guinea to the skin changes of sun-worshippers on Miami Beach. It begins by exploring the terrain of the physical skin. Picking out facts from fiction, it addresses questions such as whether our diet affects our skin, what makes our skin age, and how much sunlight is too much? These questions lead into the intriguing hinterland between the skin and the mind, from the pain and pleasure of touch to the effects of stress on the skin. The skin and the mind are intimate friends, and no other organ carries such psychological weight. How our skin is perceived by others – or, perhaps, how we think it is perceived by others – can affect our mental health. The skin is in some ways a book, in which scars, wrinkles and tattoos tell our story and can be read by others, but it is also a screen, a shifting visual display of our internal emotions, whether through subtle facial twitches, blushing or the unwanted eruptions of an underlying physical or psychological condition. The final part of this journey across our outer surface is a look at the skin in a social context. The skin unites us: humans are the only organisms that permanently mark and tattoo the skin in order to communicate with others. It also divides: skin colour and 'defiling' skin diseases have separated societies and changed the course of human history. Human skin has even influenced philosophy, religion and language, with effects that lie far beyond its mere physical form.

I hope that if you have opened this book with either a scientific curiosity or a desire for skin-health advice, you will be gratified, but that you will also be left with a grander view of yourself and of others. Indeed, this has been my journey, a wonderful adventure that began with looking at the skin on a patient or in a petri dish and ended with seeing the world altogether differently. Skin is indispensable for our survival and daily functioning, but it also reveals much about who we are as humans. The wooden écorché figures in Bologna are instantly recognizable as human forms, but without their outer covering they have lost their humanity. To understand our skin, then, is to understand our self.

1

The Swiss Army Organ

The many layers and lives of our skin

'The task is not so much to see what no one has yet seen; but to think what nobody has yet thought, about that which everybody sees.'

ERWIN SCHRÖDINGER

WE SEE SKIN all the time, both on ourselves and on others. But when was the last time you really looked at your skin? You might give it a regular inspection in the mirror, part of a daily skincare routine, but I mean properly *looked.* And *wondered.* Wondered at the elaborate, unique whorls carved on the tips of your fingers, and at the furrows and hollows of the miniature landscape on the back of your hand. Wondered at how this wafer-thin wall manages to keep your insides in and the treacherous outside out. It's scratched, squashed and stretched thousands of times a day, but it doesn't break – at least not easily – or wear out. It's battered by high-energy radiation from the sun, but stops it from ever touching our internal organs. Many of the deadliest members of the bacterial hall of fame have visited the surface of your skin, but rarely do they ever get through. Though we take it for granted, the wall the skin creates is utterly remarkable and it's constantly keeping us alive.

Never is the skin's importance more apparent than in the rare but nonetheless sobering tales of when skin fails. Thursday, 5 April 1750 was a quiet spring morning in Charles Town (now Charleston), South Carolina, but the newly ordained Reverend Oliver Hart was on his way to an emergency. Hart was an uneducated carpenter from Pennsylvania who had caught the attention of church leaders in Philadelphia and, at the age of twenty-six, was offered the role of pastor at Charles Town's First Baptist Church. (He would go on to become an influential American minister.) His diary is a humbling time capsule of the trials of eighteenth-century American life: rampant disease, hurricanes, and skirmishes with the British. In one of the diary's first entries, written a few months into his ministry, Hart records the details of that emergency morning visit, to the newborn child of a member of his congregation, because what he found was unlike anything he had ever seen before:

> It was surprising to all who beheld it, and I scarcely know how to describe it. The skin was dry and hard and seemed to be cracked in many places, somewhat resembling the scales of a fish. The mouth was large and round and open. It had no external nose, but two holes where the nose should have been. The eyes appeared to be lumps of coagulated blood, turned out, about the bigness of a plum, ghastly to behold. It had no external ears, but holes where the ears should be . . . It made a strange kind of noise, very low, which I cannot describe. It lived about eight and forty hours, and was alive when I saw it.[1]

This diary entry is the first recorded description of harlequin ichthyosis, a rare and devastating genetic skin disorder. A mutation in a single gene, called ABCA12, reduces the production of the bricks (the proteins) and mortar (the lipids) that make up the *stratum corneum*, our outermost layer of skin.[2] This abnormal development results in areas of thickened, fish-like scales (*ichthys* is the Ancient Greek word for fish) with unprotected cracks in between. Historically, infants with harlequin ichthyosis would die within days from the broken barrier as it let out the good stuff – leading to severe water loss and dehydration – and let in the bad stuff – namely, infectious agents. Without the skin's tight regulation of body temperature, the condition also poses a continual risk of either life-threatening hyperthermia or equally life-threatening hypothermia (being too hot or too cold, respectively).[3] There is still no cure for this life-shattering disease, although modern intensive treatment to repair barrier functions now enables some of these children to live into adulthood, albeit with constant medical dependence.

We so easily take for granted the countless roles our most diverse organ plays in our lives, let alone its seemingly prosaic function as a barrier. But an abnormally formed skin can be a death sentence. To begin to fathom the beauty and complexity of our largest organ, imagine hopping into a microscopic mine cart and descending through the skin's two distinct but equally important layers: the epidermis and the dermis.

The outermost layer, lying on the very edge of our body, is the epidermis (literally, 'on the dermis'). It is on average less than 1mm thick, not much thicker than this page, yet it carries out almost all the barrier functions of our skin and survives all manner of damaging encounters, which it is exposed to far more often than other body tissue. Its secret lies in its multi-layered living brickwork: keratinocyte cells. The epidermis is made up of between fifty and one hundred layers of keratinocytes, named after their structural protein, keratin. Keratin is unbelievably strong: it forms our hair and nails, as well as the unbreakable claws and horns found in the animal kingdom. The word itself comes from the Ancient Greek for horn, *keras* (from which we also get rhinoceros). If you were able to zoom in on the back of your hand to about 200x magnification you would see tough, interlocking keratin scales resembling an armadillo's armour. This biological chainmail is the culmination of the remarkable life story of the keratinocyte.

Keratinocytes are produced in the deepest, basal layer of the epidermis, the *stratum basale*, which lies directly on top of the dermis. This vanishingly-thin layer, sometimes just one cell thick,

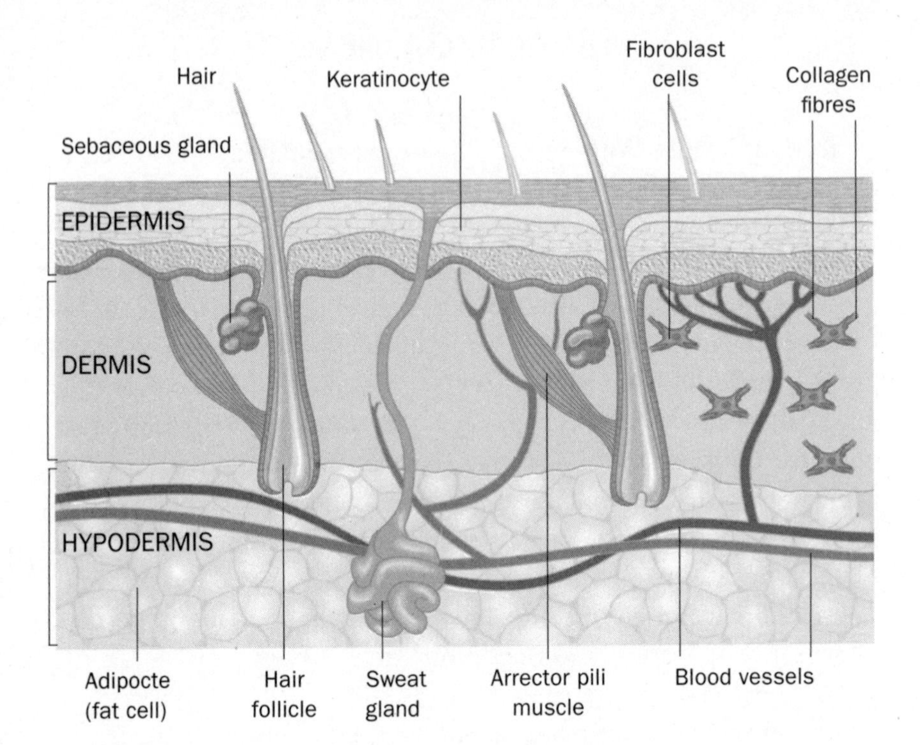

THE LAYERS OF THE SKIN

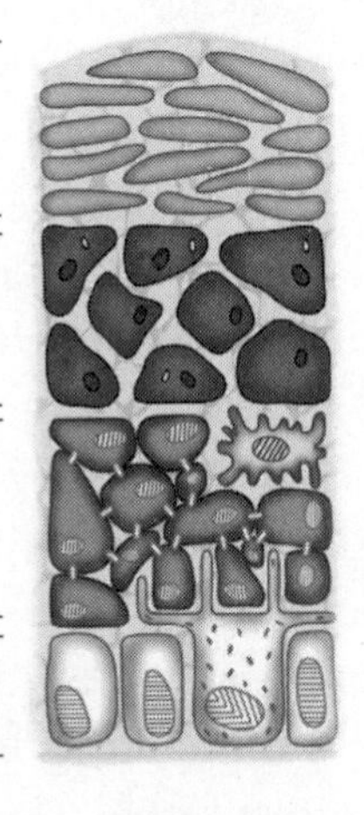

LAYERS OF THE SKIN

consists of stem cells that are continuously dividing and renewing themselves; every skin cell on our surface originally bubbled up from these mysterious springs of life. Once a new keratinocyte is created, it slowly moves upwards to the next layer, the *stratum spinosum*, or spiny layer. Here, these young adult cells start to link up with adjacent keratinocytes via intensely strong protein structures called desmosomes. They also start to synthesize different types of fat within their cell body, which will soon provide the all-important mortar in our skin's outer wall. As the keratinocytes ascend to the next layer, they make the ultimate sacrifice. In the *stratum granulosum*, or granular layer, the cells flatten, release their fats and lose their nucleus, the cell's gene-containing brain. With the exception of red blood cells and platelets, all of the cells in our body need a nucleus to function and survive, so when keratinocytes finally reach the top layer of the skin, the *stratum corneum*, they are effectively dead – but they have realized their purpose: this infinitesimally slim layer is the body's barrier wall. The living keratinocytes have become hard, interlocking plates of keratin, and the surrounding fatty mortar makes our surface as waterproof as a waxed jacket. Eventually, at the end of their month-long life, chipped away by the scratches and scrapes of the outside world, these scales flake off into the atmosphere. But this loss does not compromise the epidermal wall, with younger cells continually rising upwards to have their turn to face the world. Keratinocytes create a fine but formidable outer defence, protecting the trillions of cells inside our body. Never was so much owed by so many to so few.

An extra, fifth layer of epidermis is found in thicker areas of skin, namely the palms of the hands and the soles of the feet. The *stratum lucidum* (clear layer) is four to five cells thick and sits just beneath the *stratum corneum*. Composed of numerous dead keratinocytes containing a transparent protein called eleidin, this extra layer helps the skin at the working ends of our limbs cope with constant friction and stretching.

Coated in antimicrobial molecules and acids, the outer defences of the epidermis are chemical as well as physical, designed to keep out unwanted visitors, from insects to irritants, and to keep in moisture.[4] A watertight barrier is essential for life. In the grisly (and thankfully mostly historical) cases of humans being flayed alive, they ultimately suffered death by dehydration. When burns victims lose the majority of their skin surface, they require enormous amounts of fluids (sometimes more than 20 litres a day) to stay alive. Without our envelope of skin, we would evaporate.

The epidermis might be a wall, but it's also constantly in motion, the stem cell springs of the *stratum basale* always pumping out new skin cells. Although an individual human sheds more than a million skin cells each day, making up roughly half the dust found in our homes,[5] our entire epidermis is completely replaced each month, yet remarkably this endless state of flux doesn't cause our skin barrier to leak. This fundamental secret of the skin was discovered by means of a slightly peculiar postulation.

By 1887, Lord Kelvin, the Scottish mathematician and physicist, was already famous for his innumerable scientific discoveries, not least determining the value of temperature's absolute zero. But in his later years he sought to discover the perfect structure for foam. This odd proposal aimed to address a previously unasked mathematical question: what was the best shape to enable objects of equal volume to fill a space yet have the smallest amount of surface area between them? Although his work was dismissed as 'a pure waste of time' and 'utterly frothy' by his contemporaries, he worked his way through intense calculations, finally proposing a three-dimensional fourteen-face shape that when positioned together formed a beautiful honeycomb-like structure.[6]

The hypothetical 'tetradecahedron' doesn't exactly trip off the tongue, and for a century it seemed that Kelvin's contribution had no relevance to either material science or the natural world. Then, in 2016, scientists in Japan and London, with the help of advanced microscopy, took a closer look at the human epidermis.[7] They discovered that as our keratinocytes rise up to the *stratum granulosum*

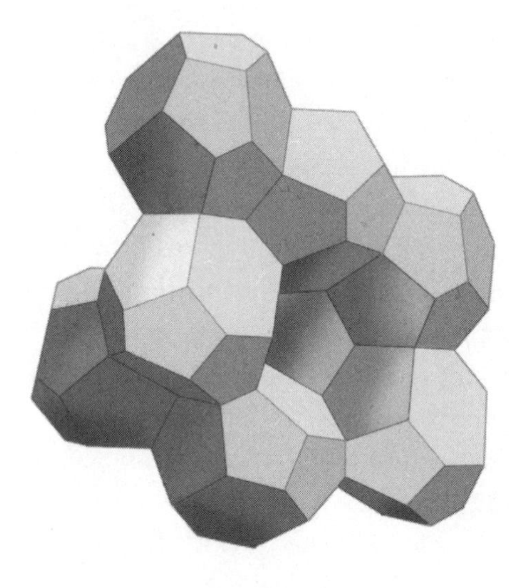

TETRADECAHEDRON

before their final ascent to the surface, they adopt this unique fourteen-face shape. So even though our skin cells are always on the move before flaking off, the surface contacts between cells are so tight and ordered that water still cannot get through. It turns out that our skin is the ideal foam. Like the intricate geometric tiles seen in medieval Islamic architecture, our skin combines function and form to make a beautiful barrier.

When our outer wall is repeatedly beaten and battered, our epidermis responds by going into overdrive, and anyone whose epidermis suffers repeated rubbing is likely to have calluses, from builders to rowers. I have a friend who, when he is indoors, will always be found strumming away on his guitar; whenever he is outdoors he climbs unfeasibly vertiginous rock faces. The wear and tear from both of these activities have prompted the keratinocytes in his epidermis to proliferate at a much higher rate than average, giving him hard, thickened calluses over his fingers and thumbs.

Callus formation – hyperkeratosis – is a healthy, protective response from our skin when it needs to reinforce the wall. But an unwanted overproduction of keratinocytes can lead to numerous skin conditions. Roughly one in three people have experienced the 'chicken skin' of keratosis pilaris, where tiny, flesh-coloured bumps cover most commonly the upper arms, thighs, back and buttocks, looking like permanent goosebumps and feeling like rough sandpaper.[8] This inherited condition is caused by an excess of keratinocytes covering up and plugging hair follicles, forcing the shaft of the hair to grow within its sealed tomb.

Keratosis pilaris is harmless and usually has little effect on quality of life, but this isn't true for all hyperkeratotic conditions. In 1731, a man called Edward Lambert was exhibited in front of the Royal Society in London. His skin (save for his face, palms and soles) bristled with black, crusty spines caused by extreme hyperkeratosis. The 'Porcupine Man', as he was dubbed, seemed to be the first of his kind. Lambert could only gain employment in a travelling circus touring Britain and Europe, and in Germany he picked up the equally undignified title of 'Krustenmann' – literally, 'Crusty Man'. He lives on in the modern name for this exceedingly rare disease: ichthyosis hystrix – *hystrix* being the Ancient Greek for porcupine.

Aside from rare genetic diseases, failures of the epidermis's all-important barrier function are also seen in more common conditions. In Europe and the USA, a fifth of children and a tenth of adults are affected by atopic dermatitis (the clinical name for eczema).[9] Ranging from mildly irritating dryness and itching to a life-ruining disease, eczema was long thought to be a purely 'inside-out' condition, with an internal imbalance in the immune system damaging the skin.[10] However, a study in 2006, led by a team at the University of Dundee, found that mutations in the gene that carries the code for the protein filaggrin were strongly associated with eczema.[11] Filaggrin is essential for the integrity of the barrier of the *stratum corneum*. It keeps the dead, interlocking keratinocytes close together as well as naturally moisturizing this layer. A loss of this protein creates cracks that weaken the wall, letting allergens and microbes from the environment into the skin and causing water to leak out. This 'outside-in' model suggests that eczema (or at least many cases of it) is caused by structural impairments in the skin barrier, rather than internal immune dysregulation. It may also explain why people with eczema experience seasonal skin changes. A study published in the *British Journal of Dermatology* in 2018 found that in the winter – in northern latitudes, at least – the amount of filaggrin produced was reduced and the cells of the *stratum corneum* shrank in the cold, reducing the effectiveness of the barrier.[12] This helps explain why eczema worsens in the cold of winter, with researchers advising extra protection with emollients over this period for those at risk. About half of people with severe eczema have a mutation in the filaggrin gene, and while it's not the only reason for this complex disease – the outside environment and internal immune system are other causes – we now know that barrier dysfunction is a prime factor.

Despite the epidermis being the most accessible part of our most accessible organ, we are still discovering its secrets. In recent years it has become evident that the epidermis is more dynamic than ever imagined. A new body of evidence suggests that skin cells contain complex internal clocks that run on a twenty-four-hour rhythm influenced by the body's 'master clock', which sits ticking away in an area of the brain called the hypothalamus.[13] Overnight, keratinocytes proliferate rapidly, preparing and protecting our outer barrier for the sunlight and scratches of the coming day. During the day, these cells then selectively switch on genes involved with protection against the sun's ultraviolet (UV) rays. A 2017 study took this one step further and found, rather remarkably, that midnight feasts could actually cause sunburn.[14] If we eat late at night, our skin's clock assumes that it must be dinner time and consequently pushes back the activation of the morning-UV-protection genes, leaving us more exposed the next day. So while studies are increasingly showing that a lack of sleep is detrimental to our overall physical and mental health, it now seems that our skin also benefits from additional sleep. The epidermis may be built to face the outside world, but it's increasingly clear that it also looks inwards, even at when we choose to eat.

Below the epidermis lies a very different layer: the dermis. The dermis makes up most of our skin's thickness, and it is a hive of diverse activity. Think of the epidermis as a factory roof, from where it's possible to look down on to a bustling workshop. Cables of nerve fibres and pipes of blood and lymph vessels snake around towering protein supports, the whole scene populated by an equally varied workforce of specialized cells.

While keratinocytes are the predominant cells of the epidermis, arguably the most important cells in the dermis are the fibroblasts – the construction workers. These cells produce proteins that act as the skin's scaffold: ropes of collagen provide skin with its strength and plumpness; elastin allows it to stretch and rebound after deformation. In between these towering structures is a gel-like matrix rich in vital molecules, such as hyaluronic acid, that carry out many other functions in our skin, including tissue repair following sun damage. The network of blood vessels in the skin is eleven miles long, enough to bridge Europe and Africa across the Strait of Gibraltar. These provide nutrients to the proliferating epidermis above and to the many specialized structures within the dermis itself.

The dermis also contains the skin's own miniature organs – sweat glands, sebaceous (oil) glands and hair follicles, which together make our skin distinctly human. If you asked any audience which characteristic has especially enabled our species to survive, thrive and ultimately dominate the planet, you might hear 'brain sophistication' or 'thumb dexterity', but the human story could never have happened without our skin's unique, if unromantic, qualities of nakedness and sweatiness.

No matter what the outside temperature, our body needs to tread an inner tightrope between 36°C and 38°C, and anything much above 42°C is lethal. The highly intelligent, but heat-sensitive, human brain could never have spread across the globe without a body capable of carrying it long distances in hot climates. This was only made possible by our industrious eccrine sweat glands. These particular sweat glands are shaped like strings of spaghetti, with one end coiled up deep in the dermis and the rest of the tube stretching all the way to the surface, opening up in a sweat pore. Our skin carries four million of these glands, and together they are capable of pumping out literally bucket-loads of sweat each day, with some humans capable of sweating three litres an hour. On a hot day, the brain's sensitive hypothalamus detects a rise in the body's core temperature and fires signals along autonomic (unconsciously acting) nerves to the eccrine glands, instructing them to send sweat to the skin's surface. When sweat – essentially water with a few trace particles of salts – lies exposed on naked skin, it rapidly evaporates. The process of evaporation removes high-energy, heat-containing molecules from the body, immediately cooling the skin and the blood vessels of the dermis. The cooled venous blood then returns from the skin to the core of the body, preventing a dangerous rise in our core temperature.

We have eccrine sweat glands all over our skin, but their density is greatest on our palms and soles. However, these areas do not seem to produce a larger volume of sweat in response to heat and exercise; instead, the glands on our hands and feet respond keenly to another stimulant of our autonomic nerves: stress. This explains why we get clammy hands as we wait outside an interview room, no matter what the temperature. Perhaps surprisingly, the sweat on our palms and soles actually increases friction and grip on the skin's surface, as our body readies itself for grappling with an enemy or fleeing up a tree. Sweat is also defensive.

But sweat is just one component of the skin's thermostat. Blood vessels in the dermis, also stimulated by the nerves, either dilate to help the body lose heat or constrict to keep it in. Compared to most mammals, human body hair is conspicuous by its absence, and our lack of it is critical for evaporation when we need to lose heat. Conversely, when we need to stay warm we might not have a thick layer of fur, but our hair follicles temporarily toil together to form another covering. The hair shafts on our skin usually lie flat, but when it's cold the arrector pili muscle attached to each hair follicle in the dermis contracts. This contraction makes the hairs erect, trapping a thin layer of warmer air above the skin and creating a temporary coat. The skin's thermostat walks a fine line, then, continuously checking and responding to our temperature to keep us alive.

Another type of sweat factory in our dermis is the apocrine gland. Apocrine glands are physically similar to eccrine glands but their product, which is oily, has served a very different purpose in the propagation of humankind; apocrine glands are found in the armpits and in the nipples and groin, hinting at their likely roles in lovemaking.

Apocrine sweat is itself odourless but its smorgasbord of proteins, steroids and lipids is a feast to the legion of bacteria on our skin, which metabolize it into the not-so-sweet smell of body odour. It has long been thought that this natural eau de parfum contains pheromones, chemical compounds that trigger a physical or social response in other humans. Although science still hasn't pinned down the exact molecules that may influence perceived attractiveness, humans are nevertheless exquisitely proficient at detecting their partner's 'odour print'. A prolonged sniff of your loved one will trigger happy memories and reduce stress levels.[15]

Apocrine sweat is also a love potion. Evidence suggests that the smell of sweat could play a role in our sexual preparedness. A 2010 study at Florida State University recruited daring (or well-remunerated) men to sniff the unwashed T-shirts of female volunteers. Intriguingly, testosterone levels increased only in the men who smelled the shirts of ovulating women.[16] This 'sweaty T-shirt' study was first devised in 1995 by the Swiss scientist Claus Wedekind, and his original experiment produced fascinating results. Forty-four male participants were asked to remain unwashed and wear the same T-shirt for two days. The T-shirts were then placed into unlabelled boxes. Forty-nine females assessed the aromas of the boxes, ranking their intensity, pleasantness and even sexiness. The results overwhelmingly showed that women were most attracted to the scents of men whose major histocompatibility complex (MHC) genes were different from their own.[17] These genes control our ability to recognize foreign molecules (and consequently dangerous microbes) and they effectively define the scope of our immune system. An individual human cannot hold a complete set of these genes; instead the countless variants are spread across the human population. This diversity means that any current or future microbe can be recognized by the immune systems of at least some humans, so a brand new flu epidemic, for example, could never wipe out all of humanity. Preferring a partner with dissimilar genes obviously makes sense from an incest-avoiding perspective, but studies also show that the offspring of partners with more dissimilar MHC genes have a more varied, and often stronger, immune system than children of partners with more similar MHC genes.[18] Our skin–nose communication, enabled by the apocrine sweat glands in our dermis, could actually be saving us from extinction.

The last of the glands in our dermis is the sebaceous gland, our skin's oil well. This small sack attached to a hair follicle secretes oily and fatty sebum up the shaft of the hair and on to the skin, lubricating both and contributing to the excellent waterproofing work carried out by the epidermis. The acids in sebum also keep the surface of the skin slightly acidic (between a pH of 4.5 and 6), which deters potentially dangerous bacteria, while those that adapt to this environment will be consequently less able to thrive if they manage to get past the skin and infect the alkaline environment of the blood. While nerves stimulate the wellsprings of our sweat glands, sex hormones have the greatest effect on sebaceous glands. This can become a problem when the pubertal rise in testosterone fuels excess sebum production, contributing to acne.

Our dermis has many tools in its inventory, and we are

continually discovering more. In 2017 researchers from the University of Cambridge and the Karolinska Institute in Sweden found that the skin of mice, and probably humans as well, helps control their blood pressure. Skin contains proteins called hypoxia-inducible factors (HIFs), which influence the constriction and dilation – and therefore the resistance – of the blood vessels in our dermis. If our skin is starved of oxygen, these proteins cause a rapid ten-minute rise in blood pressure and heart rate, followed by a drop then normalization over forty-eight hours.[19] Nine out of ten cases of high blood pressure in humans have no known cause, but some of the answers may well lie in our skin.[20]

Of the diverse cellular workforce inhabiting the skin city of the dermis, perhaps the most impressive are our immune cells. Skin is bombarded by uncountable microbes on a daily basis, which explains why it is formidably armed with an assortment of specialized immune cells. While most immune cells of the skin reside in the dermis, or are recruited there to do battle, they rely on sentinels that live up in the outer wall of the epidermis to warn them of incoming invaders. These sentinels are called Langerhans cells, discovered in 1868 by the German biologist Paul Langerhans at the tender age of twenty-one. When a potentially dangerous bacterium begins to break through the epidermis, a Langerhans cell detects the foreign invader.[21] It then engulfs small molecules from the bacterium and breaks them up into even smaller pieces. These tiny fragments, called epitopes, are unique to the particular species of bacterium. Using them rather like barcodes, the Langerhans cell then places a bacterial epitope on its surface.

What happens next is extraordinary. The Langerhans cell, holding up its captured bacterial barcode, travels far away from the skin to the body's lymph nodes. Through a series of astoundingly complicated interactions, many of which we do not yet understand, the Langerhans cell presents a 'snapshot' of the battle – showing where on the skin it is taking place and which enemy is

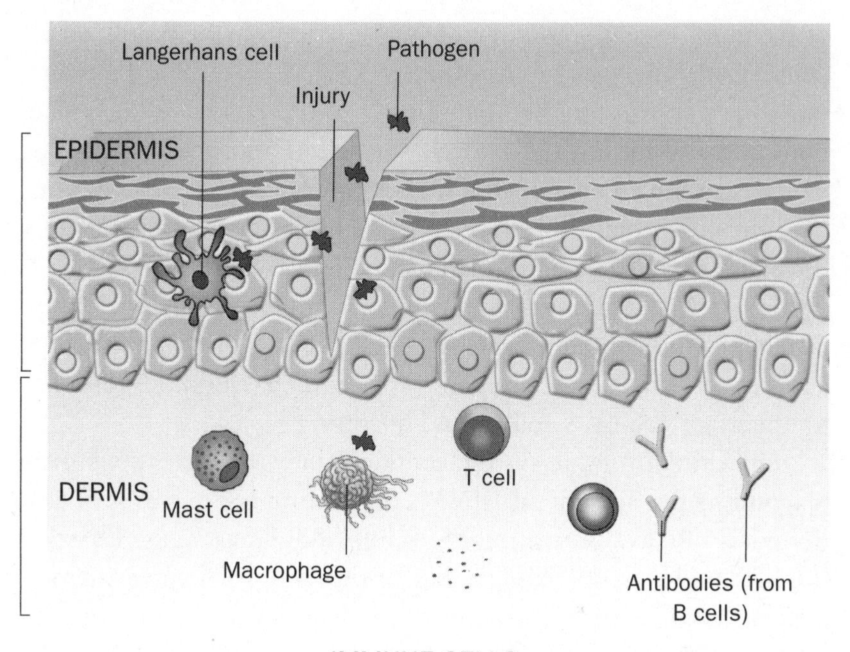

IMMUNE CELLS

involved – to a 'T cell'. T cells are able to signal to other cells and organize a coordinated immune response against any invader.[22] An even more remarkable feature of this response is that many T cells, like antibody-producing 'B cells', develop a 'memory' of this species of bacterium so that if it breaches the skin's defences in the future it can be dealt with much more swiftly.

Another example of this complex, coordinated precision-strike of our immune system is the itchy, painful rash caused by poison ivy. When a leaf of poison ivy makes contact with human skin, it leaves tiny oil molecules called urushiol that begin to travel through the epidermis and into the dermis. Some of these bind to the proteins found on the outside of skin cells. It just so happens that in almost all humans this specific oil–protein combination appears to our immune system to be a dangerous foreign microbe. In a similar process to the uptake of bacterial proteins, a Langerhans cell in the skin engulfs this oil–protein molecule, carries it to the lymph nodes deep inside the body and presents it to T cells. The first time someone's skin is touched by poison ivy, there is no allergic reaction on the skin, but the body is sensitized and ready for action. The next time an area of skin makes contact with the plant, however, the individual's T cells orchestrate a full-on attack, mistakenly believing there to be an infectious invasion. The recruited T cells destroy both the Langerhans cells containing the urushiol molecules and surrounding healthy skin cells, triggering pathways of inflammation that cause the same symptoms as the skin's response to infection: itching, swelling and blistering.

The skin's immune system contains many other weapons, each responding to a specific situation in an attempt to keep us safe. Our dermis is full of spherical, spotted cells called mast cells. They are the land mines of our skin, packed to the brim with powerful molecules, most notably histamine, which cause the symptoms of inflammation and allergy. If you want to experiment, scratch the skin along the back of your hand with a fingernail or a pointed object such as a pencil. Three things will always happen. Firstly, within a few seconds a red line appears. This is caused by mast cells emptying out their potent contents at the site, with histamine dilating the small blood vessels of the dermis to increase blood flow to the area. Secondly, after a minute or so, the redness appears to spread outside of the margins of the line. This is called the axon reflex, in which the histamine activates nerve endings, which send an impulse to the spine and back to the skin, which in turn dilates more dermal blood vessels immediately around the scratch site. Finally, a weal appears along the point of the original red line. This is because the dilation of all of these blood vessels increases their permeability, releasing blood plasma (the fluid that holds blood cells in suspension) from the vessels into the surrounding tissues. This causes the swelling that almost always accompanies inflammation. This inflammatory response is critical in our fight against injury and infection; by making all the roads to the affected area permeable, our skin enables the immune system to quickly address whatever has caused the damage.

During particularly dull lectures at medical school, one of my friends would invite me to take part in an unusual party trick. Using just a tiny amount of pressure with the end of a pencil, we would draw on his skin, which usually descended into playing noughts and crosses on his forearm. The swollen weals caused by our writing would take more than an hour to fade because he had a condition called dermographism ('skin writing') caused by excess histamine release from his mast cells. This excessive reaction is found in around 5 per cent of the global population, but the reason behind it is currently unclear.[23]

Our immune system is one of the fastest-moving frontiers of science, and skin is a fantastic laboratory. New interactions are continually being discovered; even new types of cell. In the skin immunology laboratory at Oxford University I have been able to study the roles of immune cells called 'innate lymphoid cells' in the skin, which were not even known about before the 2010s.[24] In recent years, the manipulation of our immune system using biologics – therapies that target specific immune molecules – has been turning dermatology on its head. For example, the scaly plaques of psoriasis are the result of an over-proliferating epidermis, fuelled by a dysregulated immune system. For some, these patches are just an itchy nuisance, but in visible and severe psoriasis they can be life-changing. New biologic treatments have been shown to reduce the disease in 75 per cent of sufferers.[25] With a promising pipeline of new drugs and the matching of treatments to one's genetic code, this figure is creeping ever higher and it's very likely that severe psoriasis will soon be a thing of the past.

Our epidermis and dermis are utterly distinct yet tightly connected. Both are anchored by thick, screw-like proteins to the thin 'basement membrane' that lies between them. The two layers interlock in an undulating interface, with the dermis extending up into the epidermis in a series of ridges. These ridges are most pronounced on the tips of our fingers (and toes), forming the spirals of our individuality: fingerprints. Look down at the tip of your thumb, then even closer, at the ridges and valleys that form its landscape. Unless you happen to be a member of one of the four families in the world with adermatoglyphia (a genetic lack of fingerprints), you will see one or more of the following three general patterns: whorls, which are circular, spiral shapes; loops, which start from one side of the finger, curve up and exit from the same side; and arches, which rise from one side, curve up and exit on the other side.

Our fingerprints form in the womb and are sculpted by a combination of genetic and random factors. A glance at the fingertips of a close family member will reveal the genetic component of fingerprints; their general patterns should be similar to yours. But even though some of the large patterns may appear similar within a family, the minutiae of these prints are unique; even identical twins differ. But do fingerprints have any purpose? The long-held belief that they improve grip has been questioned by studies that found that these ridges actually reduce the friction between fingers and other surfaces.[26] An alternate hypothesis is that they increase our skin's tactile sensitivity. Also, it is much harder to get

Arch

Loop

Whorl

FINGERPRINTS

blisters in these ridged areas, so they could play a role in reducing shearing forces. But, at the moment, the function of our skin's signature is still as mysterious as its individuality. What we do know is that no matter how much our fingers grow, our fingerprints remain unchanged from cradle to grave.

The fundamental importance of our literally skin-tight connection between the dermis and epidermis is sadly most evident in those who lack it. Imagine that if every time you scratched an itch or brushed your leg against a table, your skin peeled off. A penny-sized blister on a foot can be agonizing, but what if 80 per cent of your skin was a wound?

Hassan, a seven-year-old Syrian immigrant living in Germany, was born with a genetic condition called epidermolysis bullosa, in which the proteins that tightly anchor the epidermis to the dermis are missing. A shearing force as light as twisting a door handle would rip off the epidermis of his hand, causing immense pain and breaking the all-important barrier, letting water out and microbes in. The only viable skin left on Hassan's body clung to his face, left thigh and a few patches on his trunk. In this state, he didn't have long to live. Almost half of the children with this condition never make it to adolescence.

Hassan's doctors at the University Children's Hospital in Bochum, Germany, had tried the conventional treatment of skin grafting, using his father's skin, but his body rejected the foreign tissue. In 2015 they decided to seek help from Dr Michele de Luca and his team from the University of Modena and Reggio Emilia, Italy. The group had been working on some remarkable ways of developing healthy skin in the laboratory, but this had been largely untested on humans, let alone a boy with only a fifth of his skin intact. Nonetheless, cells from Hassan's surviving epidermis on his left thigh were taken and placed on a dish in the lab. Epidermolysis bullosa is caused by a mutation in a gene, called LAMB3, which is responsible for building the membrane between the epidermis and dermis, so the Italian team infected these cells with a virus that contained the healthy version of the gene, genetically modifying the cells. The team then grew nine square feet of this renewed skin in the laboratory and in two operations gave the boy's raw, wounded surface a new covering. The whole process was roughly eight months long.

Hassan's body did not reject the new skin, and for the first time in his life he had a protective outer barrier. But this was not the most remarkable discovery. When the study was published, two years after this experimental operation, Hassan's skin was still completely intact.[27] The stem cells incorporated in the new skin had formed a brand new *stratum basale*, producing fresh, healthy skin cells for perpetuity. In Hassan's landmark case, our forgotten organ was the laboratory for two emerging fields poised to revolutionize medicine: stem cell therapy and gene therapy.

As we probe deeper into the skin, it becomes hard to tell where the skin ends and the rest of the body begins. The matrix of collagen and elastin in the dermis gradually gives way to a featureless region populated by 'adipocytes', or fat cells. Whether this hinterland, known as the hypodermis (or subcutaneous tissue) is a third, distinct layer of the skin or not part of the skin at all is ultimately a matter of semantics. This unloved layer may seem fairly bland, but our adipocytes are crucial in storing energy, insulating us and providing an indispensable layer of padding. The hypodermis is also highly vascularized, making it an ideal target for injecting a payload of medication, such as insulin.

We're usually introduced to our hypodermis, however, in the form of our cottage-cheesy companion: cellulite. This upward protrusion of hypodermic fat, giving the skin a dimply, orange-peel appearance, is not a disease but a natural process in almost all post-pubescent women. Why cellulite appears in 90 per cent of women and only 10 per cent of men is all down to the architecture of the hypodermis. Subcutaneous fat is kept in place by collagen fibres that run from the dermis down to the fibrous tissue and muscles below. In women, these fibres are arranged in parallel, like the columns of a Greek temple. Prompted by a complex cocktail of hormones, genetics, age and weight gain (although cellulite is also common in young, athletic or slim women), the adipocytes can push up into the dermis, forming cellulite. Men, on the other hand, have criss-crossed collagen fibres resembling pointed Gothic arches, keeping the fat locked inside the hypodermis.

Skin is astonishing. Situated at the edge of the body, it protects us from, and connects us to, the outside world. It is familiar yet mysterious, and science is showing that the closer we look at the skin, the more we find out about ourselves. And there's still so much to explore.

2

Skin Safari

Of mites and microbiomes

'Great things are done by a series of small things brought together.'

VINCENT VAN GOGH

WHEN YOU LOOK closely at the back of your hand, it is as if you are in a passenger aircraft peering down at the world from 30,000 feet. You see ridges and canyons made of marks, scars and tendons, all dwarfed by a great mountain range of knuckles. Maybe you can make out blue rivers of veins and, if you're hairy, a forest encroaching from the arm. Just as from a plane, you can make out the terrain down below but there is no indication of any life. But when an aircraft starts to descend, you begin to detect buildings and roads, then individual cars moving along these roads. Finally, on landing and after leaving the airport, you see throngs of people on the streets, all of whom had been invisible to you from the window of the plane.

If it were possible to zoom in on the terrain of our skin in a similar way, you would enter a strange, exciting world containing diverse populations of microorganisms. Indeed, on the two square metres of our skin there are more than 1,000 different species of bacterium, not to mention fungi, viruses and mites.[1] Many of these are friendly 'commensal' bacteria, living happily on the skin neither causing harm to their host nor providing any noticeable benefit. Some bacteria are even 'mutualistic', conferring benefit to us, and are constructive components of skin society. However, others, called 'pathogenic' bacteria, are actively malicious. The lines are further blurred when it comes to 'pathobionts', a devious, two-faced type of bacteria that usually live harmlessly on the skin surface but, when circumstances change, can cause disease. This community of the good, the bad and the ugly that live with us is called the skin microbiome, and it is a complex and fascinating world. 2012 saw the publication of the first databases of the Human Microbiome Project, established to identify in detail the micro-organisms inhabiting the human surfaces, namely the skin, gut, reproductive and respiratory tracts.[2] We now know that we have at least as many – and probably more – organisms living in and on us than we have of our own cells. Tallying up the total skin microbiome is like approximating sand on the seashore, the estimates ranging between thirty-nine and one hundred trillion microorganisms, compared to our 30 trillion body cells.[3, 4] The incoming results of the project are showing that the multitudes in and upon us influence our health, and manipulating and adjusting these populations has the potential to revolutionize medicine.

In the same way that Earth has radically varied ecosystems and habitats, including oceans, deserts and rainforests, human skin has a number of habitats that support completely different populations of flora and fauna. The warm, swampy areas between the toes are utterly different from the dry, desert-like surfaces of legs. This geography is relevant to a number of diseases. For instance, the face and scalp are rich in lipid-secreting sebaceous glands – which is why they feel oily. This is a perfect habitat for the fat-loving fungus *Malassezia* and it is believed that an over-abundance of this micro-organism is the cause of seborrheic dermatitis. This strange-sounding condition is actually very common, characterized by itchy, red and flaky skin around the nose and eyebrows and dandruff on the scalp. It's often mistaken for eczema but needs to be treated differently, usually with antifungals, to remove the *Malassezia*.[5] Another disease that erupts on the slippery, oily habitat of the face is acne. Many factors cause this condition, a major one being the over-excitement of the bacteria *Cutibacterium acnes*. These rod-shaped microbes live in the dark, dingy sewers of pores and hair follicles. They feed off sebum (skin oils) and dead skin that fall down into these pores from the surface. They are usually harmless, but everything changes when the hormones of puberty kick in. With sebum secretion rising dramatically, the keratinocytes flaking off the skin surface become glued together by sebum and block the pore, forming a comedo, or spot. If this gunky mixture is completely enclosed from the outside by skin growing over it, a whitehead forms. It's a common misconception that blackheads, meanwhile, are formed by environmental dirt filling our pores, and thus indicate a lack of cleanliness. In fact, they form when dead skin cells and sebum block the top of the pore and are exposed to oxygen in the environment, causing a chemical reaction that turns the gunk a grey-black colour.

In this dark, low-oxygen environment, the *C. acnes* population booms. Its excessive breaking-down of the skin in the blocked pore causes the immune system to mount an inflammatory response and the result is the angry pimples of acne.[6] They've long been dismissed as the bottom-feeders of the skin microbiome, but a surprising 2014 study found that they have a taste for wine.[7] A form of *C. acnes* was found in the microbiome of the grapevine's stem, and it's likely that it permanently transferred from humans to the vine when we first discovered the wonders of wine roughly seven millennia ago.

One of the more unpleasant predators of our skin wildlife, present on the skin of roughly a third of us, is *Staphylococcus aureus*. Under the microscope these bacteria look like innocuous bunches of grapes (*staphyl* is Ancient Greek for grape), but they make the most of any chink in our armour. In conditions where our all-important skin barrier is compromised, such as eczema, *S. aureus* bacteria make their way through the gaps and can contribute to the pain and persistence of the inflammation.[8] They do this by releasing grenade-like toxins, such as exfoliatin, designed to break down the wall of the epidermis by damaging the anchoring proteins that hold skin cells together. In children under five this can cause the dermatological tongue-twister staphylococcal scalded skin syndrome. The toxins cause the top layer of skin to peel off, giving the appearance of a nasty burn. It looks alarming but almost always completely resolves with antibiotics. However, *S. aureus* has one more venomous trump card: enterotoxin B. If the body recognizes this toxin, the immune system lashes out and goes into overdrive. The resulting 'toxic shock syndrome' produces a sunburn-like rash, fever, low blood pressure and multiple organ failure, often leading to death. Thankfully, this disease is very rare, but *S. aureus* can still be damaging and potentially dangerous on the skin of many of us, prompting scientists to develop innovative ways of defeating it. In July 2017, Dr Eric Skaar of Vanderbilt University tweeted: 'If *S. aureus* is going to drink our blood like a vampire, let's kill it with sunlight.' His team developed a small, photoreactive molecule called 882, which activates enzymes within *S. aureus* bacteria, making them extremely sensitive to light. When the skin is exposed to a certain wavelength of light, the molecule immediately kills the bacteria.[9] This treatment is still in experimental stages, but demonstrates one of the ingenious ways in which targeting microbes could treat disease.

While *S. aureus* is overtly harmful, in many cases the threat from microbes is not so clear-cut. This is most evident in the double life of *Staphylococcus epidermidis*. These bacteria can live on our skin throughout our whole lives without causing any harm; better still, research suggests that the fatty acids produced by *S. epidermidis* actually reduce the growth of the more unpleasant bacteria such as *S. aureus*. A study from the University of California, San Diego, published in March 2018, even found that chemical compounds produced by *S. epidermidis* are able to kill some skin cancer cells whilst sparing healthy cells.[10] However, *S. epidermidis* bacteria also happen to love plastic surfaces. This sounds harmless enough, but it poses a problem in hospitals, for instance when intravenous catheters pierce the skin, picking up these little fellows and taking them into our blood vessels. Gangs of these bacteria stick to the plastic and cluster together, blanketing themselves in a comfy biofilm. This is a slimy web of proteins that anchors the bacteria to the plastic of the intravenous catheter, shielding them from both the body's immune system and antibiotics.[11] The bacterial biofilm could be life-threatening if *S. epidermidis* managed, for instance, to cling on to a prosthetic heart valve. With ongoing improvements in medical technology and surgical practice, the likelihood of these biofilms forming on prosthetic heart valves is relatively low – less than 1 per cent.[12] But if these bacteria, which are harmless on the surface of our skin, infect the inner lining of the heart – a condition known as infective endocarditis – the likelihood of a life-threatening complication becomes roughly one in two. The bacteria can even cluster together in large 'vegetations' in the heart which, when dislodged, can block the circulation to the brain and cause a stroke.[13]

Our skin's microbiome is not restricted to bacteria. Recent findings from the Berkeley Lab in California have revealed that our skin is also crawling with mysterious microorganisms called archaea. These microbes are known for being the hardiest life forms on the planet. Among them, *Pyrolobus fumarii* thrive in deep-sea hydrothermal vents at temperatures of around 113°C, and one strain even emerged unscathed from a ten-hour stint at 121°C.[14] These so-called extremophiles are so robust that space agencies actively make sure that they are not contaminants for space exploration; certain strains could almost certainly thrive on Mars. But despite their indestructible reputation, archaea are unfailingly sweet-natured when it comes to other organisms and there are no known cases of archaea causing disease in the animal world. Hoi-Ying Holman, who led the discovery of these skin microbes in 2017, believes that archaea are the caretakers of our skin.[15] Some archaea, called thaumarchaeota, could be playing important roles in the nitrogen turnover on our skin's surface by oxidizing the ammonia produced by our sweat. They may also help keep our skin acidic, making it a more hostile environment to pathogenic bacteria. Strangely, these extreme-loving organisms are found in abundance on skin at the extremes of age – in those aged younger than twelve and older than sixty – maybe suggesting that archaea avoid the oils of pubescence and prime adulthood, and prefer drier skin.

While some inhabitants of the skin are harmless-looking – most notably *S. aureus*, despite its menace – others appear so hideous we should be grateful that they are microscopic. As you read this, *Demodex* mites, which have the long tail of a worm and a body halfway between that of a spider and a crab, are most likely crawling over your face and clinging to the tree-like hair follicles of your eyebrows. At night, male *Demodex* mites come out and swim inelegantly around the surface of your face, using their eight stumpy legs to propel themselves through the skin's oil and sweat at about 16mm per hour. They are searching for a female. It sounds simple enough, but they only live for two weeks so there is a certain sense of urgency. This is on top of the fact that females live deep within the sweat glands and hair follicles, only coming to the surface every now and then to mate, before disappearing back down to lay their eggs. When these mites aren't mating, they are voraciously eating up as much sebum and dead skin as possible, but as they lack an anus this builds up inside them during their short, hectic lives, ultimately killing them. These giants of this microscopic world are usually harmless, and may play a helpful role in eating up dead tissue. However, they may also cause, or contribute to, rosacea, a common disease resulting in permanent facial redness and swelling as well as formation of nodules.[16] This is because *Demodex* mites usually have bacteria living inside them called *Bacillus oleronius*. When the mites die, often within the sebaceous glands next to hair follicles, these bacteria also die and release inflammatory proteins that stimulate an immune response, which eventually manifests as rosacea.

But despite their unsightly appearance, these creatures are also the historians of the human skin. *Demodex* mites are passed from human to human through families, potentially via breast-feeding, and a specific strain of *Demodex* can remain within a family for a number of generations, even after migration to another part of the world that harbours different strains. They are not otherwise easily transferred. This makes a particular mite's DNA a time capsule that could potentially be used to track the movement of our own ancestors across the continents.[17] Since the mites have been journeying with us for thousands of years, they may be able to tell us who we are.

While *Demodex* mites are permanent citizens of our skin, other, unwelcome, organisms frequently visit. Ectoparasites are organisms that live in or on the skin and they come in many shapes and sizes. Lice, bedbugs and fleas are tiny six-legged insects, whereas mites and ticks are eight-legged arachnids. Unless you are a budding pthirapterist (yes, louse specialists do exist) or have been infested yourself, you could be forgiven for overlooking these seed-sized parasites. The comparative giants of the skin's microscopic world

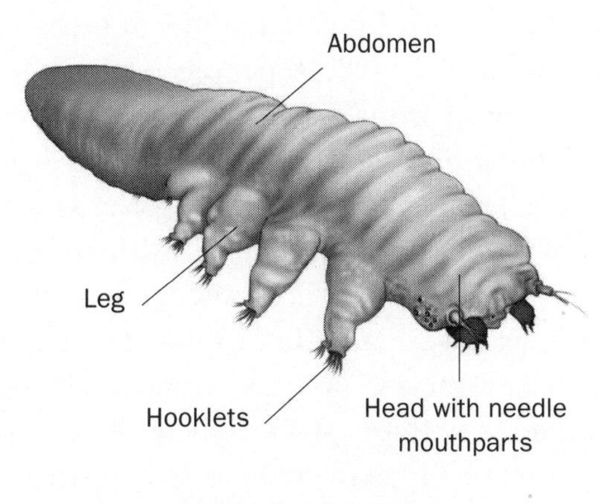

DEMODEX MITE

inhabit the different hairy regions of human skin, and each of the three species has claws specifically adapted for climbing hair of a certain thickness. The humble head louse (*Pediculus humanus capitis*) is found only on humans and spends all of its life on the scalp, growing from the size of a pinhead to that of a match-head. When a baby nymph emerges from its egg – leaving behind an empty 'nit' – and makes its first steps on to the human head, it faces a gruelling month-long life in a hostile world. Flat, elongated and wingless, these tiny insects spend their days swinging from hair to hair and shuffling around the skin's surface, having to pierce it once a day for an essential blood meal. But even its dermal dinner is fraught with danger; head lice regularly die from the pressure of human blood rupturing the lining of their gut.

When a female is ready to breed and finds a partner amongst the undergrowth, the physical strains of copulation – sometimes hours long – can be fatal. Those lucky few who survive are now able to lay a small number of eggs every day. Where along the hair the female lays her eggs depends on the outside temperature. In cold climates the female lays the egg where the hair shaft leaves the skin. If it's warm she carefully climbs right up into the canopy, as far as 15cm along the hair shaft, lays her egg and secures it to the hair by secreting a protein-rich glue.

The head louse clearly has a tough life, but its greatest enemy is its host. Up to twelve million people in the US alone,[18] and roughly 10 per cent of British schoolchildren, are infested – and the problem is not a new one: an archaeological dig at Hadrian's Wall unearthed a 2,000-year-old Roman soldier's comb, complete with a well-preserved 3mm-long louse. Although head lice are harmless, their itchy nuisance and their inaccurate association with uncleanliness have made them subject to eradication campaigns such as school 'no-nit' policies.[19] Lice can be poisoned with chemical treatments, suffocated with silicon-based lotions such as dimethicone, or manually extracted using combs.

While head lice have a universally unsavoury reputation among their hosts, who call each other 'nit-witted' or 'lousy' as an insult, some humans argue that these tiny companions may actually be our long-term allies. Humans uniquely touch and nuzzle their heads together as a sign of affection and intimacy, either romantic or familial – a behaviour absent in other primates. A team in Hungary has hypothesized that the act of touching heads is an adaptive behaviour to help share head lice with others. As the human immune system recognizes head lice as foreign, it primes itself against a potential invasion through the skin, and sharing lice generates an immune response in all members of the community. But as this insect is harmless, the immune system does not attack the head louse, but instead becomes primed to attack its deadly cousin: the body louse.[20] *Pediculus humanus corporis* looks indistinguishable from the head louse. In fact, research has revealed that they are genetically very similar and are able to breed with each other in the laboratory, leading many scientists to argue that head and body lice are actually the same species.[21] This strengthens the idea that if human skin develops an immune response to one type, it will also react to the other. In the 'wild', however, the two types of lice never venture into one other's territory, let alone mate.

Unlike head lice, body lice inhabit the sparse areas of thinner hairs across the rest of the skin and have adapted to lay their eggs in human clothing instead of hair. The most important difference for humans, though, is that body lice are the bearers of bad news, carrying, and infecting their hosts with, pathogenic bacteria. These include *Rickettsia prowazekii*, which causes typhus; *Borrelia recurrentis*, responsible for relapsing fever; and *Bartonella quintana*, the culprit of the First World War's infamous trench fever. All these infections can cause an intense fever, often accompanied by a rash.

A study in 2018 revealed that a human-to-clothes-to-human spread of body lice could also have been the primary route of transmission of the Black Death, which wiped out one third of Europe's population in the fourteenth century, challenging the traditional 'fleas on rats' theory.[22] Given that body lice thrive in squalid environments with poor hygiene and require close human contact to successfully transmit disease, their transmission is restricted to certain areas and it is unlikely that they will again be the vector for a truly global pandemic like the Black Death. But for the many who today live in impoverished and war-torn areas of the world, body lice remain a serious public health concern.

The name of the third species of human louse, *Pthirus pubis*, gives away its favoured habitat. Pubic lice inhabit the hair-dense regions of the groin, armpits, beards and eyelashes, but it is their other common name, 'crabs', that most suitably describes their squat, wide-clawed appearance, perfectly adapted for gripping thick hair. Like the other species of lice, crabs are unable to jump or fly, so they rely on intimate encounters to get from one host on to another. They don't transmit disease, but can bring with them itching, discomfort and embarrassment. While pubic lice haven't suffered national eradication attempts like head lice, they are facing a new threat to their relationship with humans: deforestation. First raised by genito-urinary doctors at Leeds General Infirmary in their well-titled article 'Did the "Brazilian" kill the pubic louse?', new evidence is suggesting that the global increase in removing pubic hair among both women and men is potentially driving the lice to extinction.[23] While many welcome the demise of this pubic parasite, Dutch biologist Kees Moeliker has instead set about collecting and storing them at the Rotterdam Museum of Natural History. He even donated one to BBC Radio 4's *The Museum of Curiosity* when making a guest appearance. Moeliker is not particularly concerned about the plight of the crab per se; rather, he hopes that his bizarre hobby draws our attention to the equally rapid destruction, and arguably more pressing issue, of worldwide animal habitats due to deforestation.

The skin's ectoparasites are not only found in the canopies of hair but also under the skin. One of my fondest memories of medical school, and a formative experience in my understanding of how to

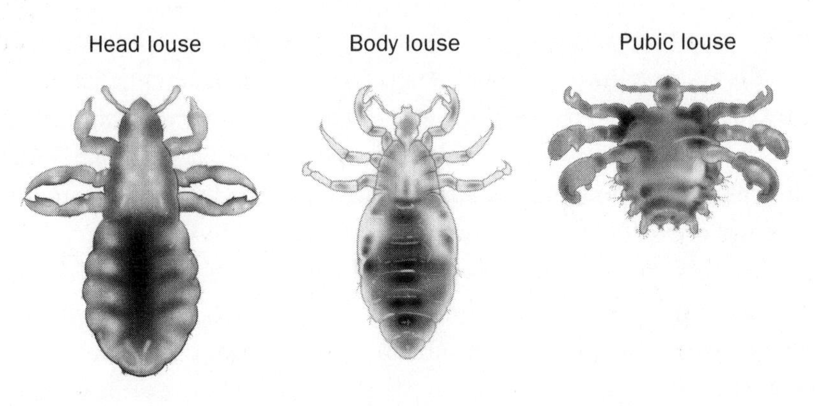

LICE

distil a diagnosis, was a period I spent shadowing an unconventional general practitioner. He was the epitome of the eccentric Englishman, complete with bow tie and mahogany pedestal desk. One of his patients, a fifty-year-old schoolteacher called Jen, entered the consultation room frenziedly scratching between her fingers. Still scratching, she sat down and placed her hands on the desk, eager for us to examine them. They appeared to be covered in a patchwork of criss-crossing lines of tiny red bumps. The doctor's eyes lit up and he brandished a magnifying glass from his inside jacket pocket.

'You may think it's eczema, but if you want to be a scabetic sleuth you must look at what's going on under the surface,' he announced. 'This looks like the work of *Sarcoptes scabiei*.' He beckoned me to look at the back of Jen's hand under the magnifying glass. I could see a small scaly patch with a tiny hole in the middle. Branching off the patch was a straight line of four or five tiny red bumps. The doctor then dipped his fountain pen into an inkwell – I assumed, for writing his clinic notes. Instead, he ran the back of the nib across Jen's hand. He then wiped off the excess ink with an alcohol wipe to reveal a blue-highlighted network of tunnels. These burrow marks are a typical sign of an infestation of the scabies mite. The female mite digs into the skin and travels along inside the epidermis, laying two to three eggs a day and forming the distinctive bumpy line as she travels. You can sometimes see a tiny black dot underneath the skin at the working end of the track, which is the mite itself. Males of the species are rather more slothful, digging shallow pits for themselves in which to rest and eat and seeking out a female's burrow when they need to mate.

Scabies' most notorious symptom is embodied in its Latin root: *scabere* – 'to scratch'. The burrows only start to itch four to six weeks after infection, when our immune system begins to recognize the allergens associated with the mite. Then, to make sure that we are better prepared for future infections, our body forms a permanent arsenal of antibodies against the mites. Next time they visit, the antibodies will rapidly bind to the allergens and the exploding mast cells will be activated within just twenty-four hours, causing a release of histamine and an intense itch. Scabies is one of the itchiest conditions known, and the infuriating and irresistible urge to scratch has led many to scrape through one or even both layers of their skin, inviting dangerous infections (*Staphylococcus aureus* often makes the most of these opportunities), and has driven some to insanity.

After hatching from their subterranean incubator, new scabies mites may roam the surface of the skin and infest others through skin-to-skin contact. These mites can also survive on towels and bedding long enough to meet a new host. Infectivity – how efficiently a pathogen can infect another host – is related to how many of these burrowing insects are living within your skin, and while most infected individuals host no more than fifteen, some populations can be drastically higher. On a visit to Australia I spoke to a dermatologist who told me about a time he was called out to visit a small Aboriginal community in the Outback. A sudden and unexplained eruption of scabies had brought the four-hundred-strong community to a painful halt. After organizing the resources needed for treatment and eradication, the dermatologist held a general health clinic. One of his last patients was a stick-thin, elderly man whose skin seemed to be completely encrusted in the silvery, thick plaques of psoriasis. On closer inspection, however, the dermatologist realized that this must be 'patient zero'. His immune system was so weak from malnourishment and old age that his skin had become a fertile field for the scabies mite. Having spent most of his life in a smaller, more remote village, he had not seen a medical specialist for decades and the population of mites on his body had probably reached one million. This infestation is called 'crusted scabies' (previously known as Norwegian scabies). It had taken only a few weeks from the old man's arrival in the village for scabies to bring its people to their knees.

But even scabies does not hold the crown for the itchiest skin parasite. This is found in a biological Russian doll: within a bacterium within a worm within a fly. In some areas of Sub-Saharan Africa, a bite from a female blackfly can be life-changing. If the fly is infected with the larvae of the *Onchocerca volvulus* roundworm, its bite releases hundreds of them deep into the lower regions of the human dermis and the fatty tissues beneath. The worm matures and mates in these deep tissues, after which the female worms can release 1,000 larvae a day into the subcutaneous tissues. At night the larvae stay under the skin, but daylight draws them into the upper layers, ready to be eaten up by female blackflies, which only feed during the daytime.

Hundreds of these larvae never end up catching a flight inside one of these insects, however, and die en masse in the skin.[24] The death of an *Onchocerca* larva releases the bacterial members of its own microbiome into the skin of its human host, but 'good bacteria' to a worm are not always beneficial to us. Our skin's immune cells immediately recognize *Wolbachia* bacteria, which commonly live inside the worms, and the resulting inflammation brings about the severe itch of 'onchocerciasis'. The ability of the larvae to infect the eye and damage eyesight has given the disease its more common name of 'river blindness'. When visiting East Africa, I often heard stories of sufferers scraping their skin incessantly for weeks on end, scratching through to the muscle using anything to hand: fingernails, broken pottery, even machetes. No deaths have been directly attributed to the *Onchocerca* worm, but the physical and social complications of unstoppable itching knocks thirteen years off the average life expectancy of anyone infected.[25]

Our relationships with the bugs that live on our skin have opened up some unexpected avenues for medical treatment. The house dust mite sounds harmless, but it is a deceptively menacing creature. It lives in the warm, damp environments of bedding and furniture and its favourite food is dead human skin – given an opportunity, dust mites rush on to our surface, causing or contributing to skin irritation and eczema. Researchers at the laboratory I worked in at the University of Oxford discovered that this is due to a substance these mites produce, called phospholipase, which breaks down fat molecules in the skin that consequently stimulate the immune system. This amplifies the immune response at the affected area of skin, causing the red, itchy inflammation of eczema.[26] This is one of many examples of how new research into skin microbes may open up new avenues for treatment.

Out of all the organisms that reside on and visit our skin, tick spit would not intuitively appear to be a goldmine of revolutionary medical treatments. When a hungry *Ixodes scapularis* (black-legged deer tick) latches on to human skin, it plunges its hard head down into the dermis in search of a blood meal. Some of these ticks harbour the spiral-shaped *Borrelia* bacteria in their gut, taken from previous acts of vampirism on other mammals. When these bacteria enter human skin from the infected tick, they begin to migrate across the skin in all directions, creating a red rash. Like the ripple of a pebble dropped on to a still pond, the rash migrates outwards in an expanding ring, leaving an area of central clearance. This distinctive bull's-eye rash is a result of our inflammatory response frantically trying to catch up with the fleeing microbes. Termed erythema chronicum migrans (chronic migrating redness), this characteristic rash is enough to diagnose

Lyme disease: a condition of fevers, agonizing joint pain, memory loss and heart palpitations. The reason for the bacteria's success in spreading around the body lies in the tick's saliva: a mysterious potion abounding in molecules that suppress our immune response. The many-thousand unique proteins in the saliva neutralize our immune molecules and trick our immune cells, enabling the tick to feed for more than a week without our body even registering its presence. Scientists realized that if the tick's saliva can suppress our immune system, some of its molecules could also be used to dampen down unwanted inflammation and treat autoimmune diseases. In 2017, a team from Oxford University isolated one protein – with the rather unmemorable name of P991_AMBCA – from the tick's salivary treasure trove and found that it binds to, and inhibits, certain chemicals released during myocarditis, a potentially fatal heart disease.[27] Whether microbes or other tiny organisms live on our skin throughout our life or opportunistically visit for a week, modern science is finding new ways of using their abilities for our wider benefit – from bug to drug.

The make-up of our skin's microbiome also influences the health of our skin, and manipulating it could change the way we treat skin conditions. Starting at the very beginning of life, there is evidence that even the way in which we are born, whether vaginally or by caesarean section, could determine our future skin and gut microbiomes. When we first enter the world, as a greasy, squealing baby, our skin is largely a blank canvas that is ripe for colonization and immediately some of the microbes living in either the mother's vagina or the skin around the C-section incision, as well as those in the hospital environment, will decide to make their home on the newborn baby's surface. Exactly which types of pioneer species land on an infant's skin is important and can have long-term consequences, as these rapidly become the dominant organisms of the skin, making it hard for other bacteria to later gain a foothold.[28] Vaginal microbes contain more 'good' bacteria than the mother's abdominal skin and those of the hospital environment, which has more of the pesky *S. aureus*, and this may be why C-section babies appear to have an increased risk of developing allergies later on in life. So should we start swabbing all newborn babies with their mother's vaginal mucus? This process, termed 'vaginal seeding', is not yet mainstream but is starting to increase in popularity; in Denmark, for example, 90 per cent of obstetricians have been asked about vaginal seeding by expectant mothers.[29] At the moment, however, there just isn't enough scientific evidence to support the treatment. The long-term effects of vaginal seeding aren't known, and some studies argue that children born by C-section may have an increased rate of allergies due to factors associated with the need for the procedure, such as conditions requiring the mother to take antibiotics.[30]

A few days after we are born, there appears to be a window in which our skin becomes populated by 'regulatory T cells'.[31] These influence the development of other immune cells, helping dampen down excessive responses to bacteria and stopping immune cells attacking 'self' molecules resulting in autoimmune disease. The type of microbes that end up on our skin at the start of our lives may affect how these and other immune cells develop. Although we don't yet know the effects on disease later in life, this early skin-programming potentially has knock-on effects across our immune networks, impacting functions in organ systems such as the gut and the brain. It's essential that we have a healthy, mutual relationship between our immune cells and the bacteria that inhabit the niches of our skin, otherwise our outer surface would be in a constant state of turmoil. Patients with immunodeficiencies – where they lack many of the combatants of our body's immune force – tend to have more diverse populations of bacteria because their skin has become too permissive without the border control of immune cells.[32] Such discoveries are opening up intriguing questions, such as whether a mother's diet and antibiotic use before and during pregnancy affects the baby's immune system development and its subsequent microbiome.

Given how exposed these tiny microbes are on our surface, one might reasonably expect that over time they would be blown off by the wind, drained away with our sweat or lifted off into the air with the million skin cells we shed each day. Interestingly, a team from the Jackson Laboratory for Genomic Medicine in Connecticut has found that, despite our skin's exposure to the external environment, its microbial populations remain largely stable over time.[33] Our hands – areas one would assume to be only a temporary crossing point for migrating microbes, which are either washed away under the tap or swapped in a handshake – actually have a high level of stability. It's easy to imagine the bacteria as bugs living on a flat layer of skin, but in fact they are tiny organisms, thousandths of a millimetre long, hiding in the canyons and crevices of our surface. The populations of microbes on our skin do change over time, but they are not clinging on for dear life every time we have a shower.

The main change to our skin's microbiome happens when we go through puberty, when the proliferation of sebaceous glands creates an oily surface that favours fat-loving bacteria, such as *Propionibacteriaceae*, which start to replace the usual populations of commensal skin bacteria and set the stage for acne. Our skin microbiome largely stabilizes during adulthood, giving us a kind of microbial signature, but that does not mean there is no microbial migration. An unconventional study at the University of Oregon in 2013 explored this by closely examining roller derby players. This contact sport involves two teams racing on roller-skates around a flat, indoor track, with constant blocking and scrimmaging in an attempt to overlap the opposing team. The study found that frequent skin-to-skin contact during training resulted in members of a roller derby team sharing similar patterns of skin microbes.[34] But when two sides clashed during a match, both teams left with different patterns of bacteria, having gained additions from the opposing side.

Whether we like it or not, when we start living in close quarters with someone, we also begin to share their microbiome.[35] A 2017 study found that cohabiting sexual partners could be identified from a group of random individuals in nine out of ten cases, based solely on their skin microbiome profiles.[36] This adds a slightly uncomfortable dimension to the marriage notion that 'two become one flesh'. The study found that couples shared most microbial similarity on their feet, and had least in common on their thighs. In the study, the biological sex of each of the participants could be determined simply from thigh microbiome samples, most likely due to distinct groups of microbial travellers from the vaginal microbiome. These household microbial signatures can be expanded to whole cities: one study measured the microbial make-up of various offices – including the skin of the office workers – across three separate cities in North America: Flagstaff, Arizona; San Diego, California; and Toronto, Canada.[37] Intriguingly, each city left its own microbial signature on its employees, even across different offices in the city, so the city where an employee lived and worked could be determined simply by examining their skin microbiome. When you think of the millions of microbes that cling to the flakes of our shedding skin in a confined office, or the number of hands that grab on to a pole on a subway train, you can see why we have more in common with our neighbours than we think. In London it's almost anathema to speak to a fellow passenger on the Underground, but we may already be sharing more than we'd like.

One cold winter's evening a few years ago, I was sitting in the corner of a small Oxford pub with a group of immunologists. I brought up the subject of the skin microbiome and one friend, who always seems to ask the right questions, asked: 'So, we can share our microbiomes with others. If I have a particularly bad composition of bacteria on my skin that's worsening my eczema, could I slowly "infect" my partner with it?' Not long afterwards, in 2017, a team from the University of Pennsylvania started to answer this question.[38] They infected the skin of a mouse with the parasite *Leishmania*, which altered its skin microbiome. This pathogenically altered microbiome was then passed on to the other mice in the cage, even though they were never infected with *Leishmania*. Although still barely understood, the strange, fluctuating world of the skin microbiome is slowly being revealed, and it is changing how we view life.

In the developed world we have incredibly high levels of hygiene compared with our ancestors, and children are far less exposed to infectious agents than they would have been a century ago. This is great news when it comes to infectious diseases, but a lack of exposure to bacteria in early life may in fact impair the normal development of the immune system, particularly 'immune tolerance'.[39] This is the ability of our immune system to hold itself back and remain unresponsive to substances that are harmless or belong to our own body. The potential result of this stunted development is that our skin's immune system becomes hyperactive and skewed towards allergic and inflammatory conditions. This 'hygiene hypothesis' is a compelling explanation for the high rates of eczema, hay fever and asthma in the developed world.[40] So how can we treat conditions, such as eczema, that can be worsened by this lack of bacterial diversity and reduced immune tolerance? Some of us may have been literally bathing in the answer.

Since time immemorial humans have flocked to thermal baths to find health and happiness, with resort towns and even whole cities developing beside the springs. One of the most famous is the city of Bath in south-west England. It's fascinating to think that 2,000 years ago Roman citizens and soldiers would come here for rest and recuperation after a day's work or a season of hard fighting against naked British savages, reclining and scraping off dirt and sweat with olive oil and a *strigil*, a sharp, metal scraping tool. Minerals in spa waters have long been touted as having healing properties for skin diseases, but recent discoveries suggest that this could be the result of microbes that have been bathing in the waters since long before the Romans. Among these are *Vitreoscilla filiformis* bacteria – tiny, see-through organisms that glide across the surface of the skin and have been shown to reduce inflammation in eczema. A 2014 study found that *V. filiformis* talks to our immune system through a series of signalling pathways, which in turn produces more regulatory T cells, dampening down the immune response and helping to alleviate eczema.[41] Right now we still largely rely on steroid creams to dull the immune response of eczema; creams containing bacteria such as *V. filiformis* could one day provide a sustainable, side-effect-free alternative.

In the past decade, the gut probiotic market has increased exponentially. Probiotics are live bacteria that are the same as – or similar to – those that dwell within our bodies. Today millions of people take their daily dose of yoghurt that has been laced with 'good bacteria'. The science behind adding populations of bacteria to the gut microbiome to improve our health is mixed, but good logic stands behind it. Infections with the bacterium *Clostridium difficile* affect the gastrointestinal tract, producing watery diarrhoea and abdominal pain, and can lead to the life-threatening complications of bowel perforation or sepsis. Faecal microbiota transplants – which effectively involve eating the freeze-dried poo of healthy donors to displace that pathogenic bacteria with harmless ones – have recently been shown to be incredibly effective for those with a *C. difficile* infection in their gut.[42] In many ways the skin also lends itself to this kind of therapy, helped by the fact that, unlike the gut, the skin has no acidic stomach that can kill the bacteria.[43] One example currently under investigation is the application of *Staphylococcus epidermidis* and *Staphylococcus hominis* to the skin to displace populations of harmful *Staphylococcus aureus* in diseases such as eczema.[44] Meanwhile, new research is revealing that teenagers with acne have a more diverse skin microbiome than those without the condition, so adjustments to the microbiome could also have a beneficial effect here.[45] It's not inconceivable that microbiome 'transplants' could soon be used on our skin.

Probiotic skin treatments could even be the answer to body odour. Apocrine sweat glands populate our armpits, genitals and nipples. Unlike the eccrine sweat glands across the rest of our skin, which produce watery sweat as long as necessary to cool us down, apocrine glands produce large amounts of oily sweat in short bursts. This oily sweat is completely odourless, but bacteria, specifically *Corynebacteria*, break up the oils into foul-smelling molecules. One of these is butyric acid, which gets its name from rancid butter (in which butyric acid was first discovered). This chemical gives human vomit its distinctive smell, which is so powerful our noses can detect it even if it makes up just 0.001 per cent of the molecules in a single sniff. The potency of this combination of apocrine sweat glands and *Corynebacteria* is demonstrated in those who have few of either. East Asians, especially Koreans, have significantly less body odour than the rest of the world, due to a genetic make-up that produces fewer apocrine sweat glands and favours different populations of underarm bacteria.[46] For those who suffer the mental and social effects of strong body odour, spreading our armpits with probiotics could one day be the solution.

At the 2017 Karolinska Dermatology Symposium in Stockholm, Sweden, the attendants were introduced to the results of a world first: an underarm bacteria transplant.[47] For his experiment, Dr Chris Callewaert of the University of California, San Diego found a pair of identical twins, one of whom didn't seem to smell at all while the other had particularly strong body odour. He asked the odourless twin not to wash for four days so that he could cultivate a good quantity of whatever species of sweet-smelling bacteria he possessed, whilst the other had to scrub his armpits each day for the four days to prepare his skin for the new population of microbes. The dead skin of the first twin was scraped off and swabbed into the armpits of his smelly double. Remarkably, the smelly twin's odour disappeared and the effects lasted for a year. Although these are still early days, the promising results were repeated in sixteen out of eighteen further pairs. Perhaps we will soon be ditching deodorant in favour of donations from unscented friends.

At first glance, our skin looks like a bare, inhospitable landscape. It's clear, however, that our body is covered in habitats filled with wildlife worthy of a nature documentary. Our skin is their world. We are continuing to deepen our understanding of how the skin's microbiome causes and influences skin disease, but it is already apparent that adjusting the balance of microbes on our surface could be an answer to a number of skin problems.

Facing out towards the environment, skin is the most easily accessible laboratory on our body, a bonus given that skin microbiome research could be of great use in the study of diseases affecting other organs and body systems. In turn, as we are also about to see, ripples from microbiomes in seemingly distant organs can also have a direct effect on the skin.

3

Gut Feeling

The relationship between our insides and outsides

'Tell me what you eat, and I will tell you who you are.'
JEAN ANTHELME BRILLAT-SAVARIN

IN MY FIRST days of hospital placement during medical school, we were shown how to perform abdominal examinations. This involves looking for signs of diseases associated with the gastro-intestinal tract and liver as well as feeling, tapping and listening one's way around the gut. Long before we got around to feeling for pain and abnormalities in the patient's abdomen, however, we were told to look closely at the skin, to see what it might be able to tell us. Dozens of changes to our outer covering can help build a picture of what is going on underneath. This might be the yellow of jaundice, the reddening of the palms in liver disease or the strange black patches across the armpits that may indicate stomach cancer. I was most fascinated by the 'spider naevus', a central splodge of red with web-like radiating vessels found on the chest and back of patients with liver disease. A simple but dramatic test distinguishes them from other red skin marks: a gentle press and release of the central spot triggers an ink-run of blood to refill the empty veins. I was captivated by the idea of 'reading' the skin, communicating with messengers who told stories of distant and unseen internal organs.

We all instinctively feel that what happens in our insides has some effect on our outsides. We often feel that our diet affects the health and appearance of our skin, from an overindulgence in Easter eggs causing that nasty outbreak of spots to our complexion being improved by drinking more water. Although the skin and the gut are two completely different continents, science is slowly revealing that they do indeed communicate with each other, through a network of hugely varied and largely undiscovered silk roads. Some paths are direct, seen in the eruption of weals and rashes on the skin from eating a food allergen. Others, such as the effect of a healthy diet on the skin, are winding and controversial routes, guided by the trade winds of genetics and environmental factors. Answering the seemingly simple question, 'Does diet affect my skin?' requires wading through a swamp of scientific – and some not-so-scientific – studies and conflicting literature, while the complexity of the human body means that even conclusive laboratory findings do not always translate into human studies. All the while, public (and sometimes professional) opinion is affected by celebrity endorsements, fad diets and the influences of the food and pharmaceutical industries, both eager to sell their products. It is no surprise that diet divides dermatologists. This complex frontier of medicine not only demonstrates the successes and failings of science, but also – and more importantly – the wondrous complexity of the relationship between our skin and our gut.

Lying low to the Pacific Ocean, roughly 100 miles from mainland Papua New Guinea, sits a tiny tropical paradise. Covering less than ten square miles and inhabited by just over 2,000 islanders, few people have ever heard of Kitava. This unassuming island was chosen by the Swedish professor Staffan Lindeberg and his team to be the site of a landmark study. At the time of the study, Kitavans were one of the last peoples on the planet completely untouched by a Western diet. Almost all lived off an intake of fruit, root vegetables (such as yam and sweet potatoes), coconut and fish. It was predominantly plant-based, with a high-carbohydrate intake of low-glycaemic-index foods, but it was not low in fat. Alongside the somewhat disputed findings of low incidences of heart disease and stroke – which helped kick-start the so-called 'Palaeolithic diet' – one of the most startling results of Lindeberg's study was the complete absence of acne in all 1,200 subjects: 'Not a single papule, pustule, or open comedone was observed in the entire population examined'. This raised the possibility that something in the Western diet could be at least partly responsible for skin disease.

Seeming to support the Kitava findings, it is apparent that when diets become Westernized, the levels of acne increase. The foods with the strongest evidence for causing and contributing to acne are those with a high glycaemic index, which quickly and significantly raise blood-sugar levels. Even when weight, age and sex are accounted for, the incidence of acne is greater among those with higher intakes of high-glycaemic-index foods.[1] Sugary foods and certain carbohydrates cause a spike in insulin and insulin-like growth factor-1 (IGF-1), both of which inhibit a gene regulator called FOXO1. This process has a number of effects on the skin: fat synthesis within the skin increases, sebum-producing cells proliferate and the skin loses the ability to control levels of the bacterium *Cutibacterium acnes*.[2] Milk also contains IGF-1 plus dihydrotestosterone and growth factors, leading some studies to suggest that milk intake is associated with acne development. Low-fat milk could be particularly to blame, as one theory suggests the responsible hormones are less diluted by fat. The evidence for milk causing acne, however, is not as strong as that for high-glycaemic-index foods. British dermatologist Stefanie Williams reflects a growing consensus about diet's role in the health of our skin – not just in relation to acne – when she says, 'The current Western low-fat obsession, with over-reliance on starchy, grain-based and sugary foods, doesn't do our skin any favours.'[3]

'Don't eat too much chocolate; it'll give you spots!' This daily warning given by parents to their children is a classic example of a common but misplaced assumption about a specific skin–gut relationship. Most scientific evidence suggests that chocolate does not have a significant effect on acne. This sugary snack actually has a low glycaemic index due to its high fat content, which slows down sugar absorption. Interestingly, however, much of the evidence for chocolate not being responsible for spots comes from a 1969 study that remained completely unchallenged for four decades.[4] The many questions surrounding the integrity of the study (namely, that it was funded by the Chocolate Manufacturers Association of the USA) have led to a recent revisiting of the subject.[5] While one study found that a binge consumption of 100 per cent cocoa did seem to exacerbate acne in males, only thirteen participants were studied.[6] It is really important to note that small sample sizes in studies have to be taken with particular caution, as they are unlikely to represent the true variation we see in the whole population. But if there currently isn't enough evidence to implicate chocolate as a pustular villain, why is it always being portrayed as the perpetrator? It could be down to a classic confusion of cause with correlation. An interesting answer could be that, on average, women tend to crave sweet treats during the premenstrual part of their cycle, which is just when androgen levels rise and acne breakouts increase. The correlation – my spots come back when I eat chocolate – is not the same as the cause: hormonal changes around the menstrual cycle are proven to be the most likely reason for the spots.

In the case of acne, it is evident that our gut can communicate with our skin via metabolic and hormonal changes. But this doesn't address whether food particles can directly reach and influence the skin. Britain's love of the South Asian curry is no secret, and it is claimed that chicken tikka masala is the country's most popular dish. I used to regularly visit the 'Balti Triangle' area of Birmingham – the birthplace of the Balti curry – with a spice-loving friend. He claimed that after his weekly chicken jalfrezi, a slight garlicky smell he called 'Eau du Balti' lingered on his skin for a couple of days. He argued that the smell comes from sweating out food molecules from the curry, but I was convinced that any odour must have been down to the aromas wafting directly from the plate and sticking to his clothes and skin.

It's certainly true that spicy foods make you sweat: the capsaicin molecule in chillies triggers exactly the same receptors on the tongue and skin that are activated when either of these tissues are burned. This tricks the brain into thinking that we are hot, so the body tries to cool us down by sweating. But can food be excreted in our sweat?

It turns out my friend was right: while the vast majority of food waste ends up in faeces, some volatile, aromatic compounds are released in breath (e.g. allyl methyl sulphide molecules of garlic) and urine (such as the asparagus smell experienced by roughly half of the population) and, indeed, some individual molecules can be detected in sweat. Some of the biggest sulphur-containing culprits include garlic and onions – sulphur is not only responsible for the 'rotten egg' smell as commonly believed – but interestingly the same molecule that generates bad breath appears to have a different effect on the smell of our skin. Scientists from the University of Stirling and Charles University in Prague discovered that, on average, women found that the body odour of men who had eaten 12g of garlic was more attractive than those who had eaten either 6g or none.[7] In another sniffing study, females preferred the aroma of sweat from vegetarian men compared to that of meat-eaters.[8] I salute the inventive scientists who designed experiments like these, but not as much as their brave, sweat-smelling participants.

For some people, however, the food molecules that reach their skin can have life-ruining consequences. A colleague of mine who works as a general practitioner will never forget the day he met Sally. As soon as this patient – a fit, healthy woman in her mid twenties – walked through the door, the doctor was almost knocked back off his chair by her pungent smell, which could only be described as a wall of rotting fish. Over the previous two years Sally had begun to notice office co-workers shifting their workspaces further away from her desk, pedestrians screwing up their faces as she passed and teenagers holding their noses and pretending to retch as she stood next to them at the bus stop. Despite washing twice a day and slathering on perfume, the last straw came the week before the consultation, when Sally was asked by a waiter to leave a restaurant and 'have more consideration for the other diners'. The GP, who admitted he had no idea what was going on, referred Sally to hospital, where genetic testing revealed the cause of her odorous ordeals. She had the rare genetic condition trimethylaminuria, or 'fish odour syndrome'. Trimethylamine is a compound synthesized by gut bacteria breaking down certain foods, including fish, eggs, beef, liver and specific vegetables. People with trimethylaminuria lack the enzyme that breaks this molecule down, so the excess trimethylamine makes its way into the sweat to create a rubbish-dump cocktail of rotting fish and eggs. A strict change of diet almost completely eliminated Sally's stink, and it gave her back her life.

If it is possible for certain molecules from food to affect, and indeed travel to, the skin, it's unsurprising that many believe that nutrition can directly affect the skin's health, including reducing the risk of developing skin cancer. However, again we sink into the swamp of conflicting evidence, and there are few clear answers. It is a widely held – but scientifically contested – belief that 'antioxidant' supplements (molecules that are supposed to inhibit damaging oxidation reactions in our cells) reduce our risk of cancer. However, there is nothing to concretely show that antioxidants, including β-carotene and vitamins A, C and E, reduce the risk of skin cancer in humans.[9] In fact some, such as selenium supplements, may even increase the risk of cancer if taken in high doses. A few have been shown to reach our skin after eating, such as the catechins found in green tea, but the jury is still out as to whether these have any beneficial effect.[10] Any effectiveness of antioxidants in laboratory studies cannot currently be repeated in humans due to the way these molecules are broken down and metabolized. But even with limited evidence and lacking any scientific consensus, antioxidants are incredibly popular and are ubiquitous in advertisements for healthy foods. So why is this the case? As humans, we are hard-wired to find the route of least resistance to achieving safety and good health, and the idea of individual 'silver bullet' foods is very attractive compared to what has actually been shown to work: a balanced diet including fruit and vegetables, plus regular exercise, not smoking and limiting alcohol intake. The multi-billion-dollar health-food market exploits this desire for easy answers, which really are too good to be true.

This is not to say, however, that none of these molecules confers any benefits; it is possible we do not yet know about them, and it is very likely that other components of antioxidant-rich foods could aid our health. For instance, retinol, a form of vitamin A found in eggs and dairy products, has been shown to decrease the risk of non-melanoma skin cancer for those at moderate risk.[11] Lycopene, a carotenoid found in tomatoes, has also gained a considerable degree of attention as it appears to halve the rate of skin cancer risk in mice.[12] Interestingly, tomatoes themselves have a stronger effect than pure lycopene, suggesting that other components of these fruits may be beneficial. Whether these red fruits reduce human skin cancer is yet to be seen, but there is an additional bonus to eating colourful foods high in carotenoids, such as carrots, tomatoes and peppers: these foods give you glowing skin. For those seeking a 'healthy' tan, diets supported by food rich in carotenoid pigments have also been shown to induce a small but noticeable golden glow.[13] In experiments where subjects have to measure the 'attractiveness' of white-skinned participants, those who ate plenty of fruit and vegetables had similar ratings to a mild suntan, without any of the negative effects of UV radiation.[14]

Other studies show even more remarkable results. Dr Ian Stephen from Nottingham University, and others since, showed that when women are asked to judge men's faces – of any background skin colour – those with a yellow-red glow from carotenoid foods were deemed more attractive than those with either paler skin or a suntan.[15] The golden-glow effect was an even more attractive magnet than the masculinity of the face. Perhaps this isn't too surprising, considering the logic behind finding a sexual partner: red-yellow skin suggests a healthy individual with a strong immune system, and we are more likely to gravitate towards a person who looks like this and potentially mate with them. Sometimes it's easy to spot changes in health through the skin, whether that's the pale skin of someone with severe anaemia or the blue-tinged extremities of a poorly oxygenated sick patient. But as humans we are also remarkably adept at picking up on very subtle changes in the skin of others, even changes we cannot describe or express.

Professor James Watson – the co-discoverer of the DNA double-helix – has a point when he criticizes eating specific foods for their supposed antioxidant properties: 'Blueberries best be eaten because they taste good, not because their consumption will lead to less cancer.'[16] This comment has to be received in the knowledge that there is a clear benefit of a balanced diet for healthy skin. It is certainly worth eating lots of fruit and vegetables, but it is not worth pursuing one particular 'superfood'.

Eczema is another example of the complexity of skin's relationship with diet. I have seen patients who take some (and occasionally all) of the natural supplements believed to help alleviate the condition. One thirty-year-old woman with severe eczema had eschewed any medication or cream that she deemed 'medical'; instead, at home she seemed to be drip-feeding herself primrose, borage, sunflower, buckthorn, hempseed and fish oils, as well as taking zinc sulphate and selenium tablets. None of these seemed to be working. For every patient who appears cured by adding any of these items to their diet, there are dozens for whom it has not worked at all and the symptoms of eczema were instead relieved by conventional treatments. While food allergies can exacerbate eczema, there is currently a lack of real evidence that any specific foods significantly reduce the effects of this common, and often debilitating, disease.[17] This is not to say there aren't any foods that alleviate the symptoms of eczema in anyone; given the huge variation in our genetic and environmental make-up, it seems highly possible that some people will discover that certain foods seem to help their particular condition. However, foods that seem to work on a case-by-case basis should be tried alongside – not instead of – conventional treatments.

Supplements for skin diseases such as eczema highlight one of the difficulties in nutritional medicine, and in medicine as a whole: the human body is much more complicated than we would like it to be. Slightly more promise has been seen in the use of vitamin D to improve eczema symptoms. Produced in the skin in response to ultraviolet rays from the sun, vitamin D has direct effects on mineral absorption in the gut. In this way, our skin and gut work together to strengthen our bones and immune system, and in one trial, people taking 1,600 IU vitamin D a day had improved eczema symptoms, but subsequent studies had different results.[18] Vitamin D has, however, been found to improve eczema in a subset of patients who had recurrent bacterial infections, suggesting that this molecule plays a role in our immune system.[19] Even though we are uncertain about the effects of vitamin D on the health of the skin itself (see Chapter 4), supplementation is a safe, health-service-endorsed route to prevent deficiency of this vital vitamin for the rest of the body, particularly in winter months in countries far from the equator. One potential breakthrough with eczema, however, has been the discovery in 2017 that a large subset of eczema patients has a mutation in the CARD11 gene, which is involved in the activation of immune cells in the skin.[20] The molecule glutamine appears to remedy the effects of this mutation, so researchers are planning to see whether diet supplements of glutamine could improve this disease.

If eczema research leaves us with the sober realization that there may be few wonder-cures in our food, the example of psoriasis directly encourages a healthy, balanced diet. There is significant evidence to show that weight loss greatly improves psoriasis while obesity exacerbates its visible, scaly eruptions.[21] In obesity, the human body is in a pro-inflammatory state, so weight gain increases outbreaks of psoriasis. Being visible, psoriasis tends to push people into circles of despair, with the disease itself contributing to obesity by encouraging social isolation and bad eating habits, which further exacerbates the condition. Psoriasis is also made worse by alcohol, which can also lead to sufferers forming unhealthy dietary habits and avoiding medical help just when they most need it.

Another little-explored story of the skin and the gut is that almost a quarter of psoriasis patients have gluten sensitivity and – although the exact mechanism is not yet known – in some patients with this scaly skin disease, a gluten-free diet can significantly help treat it. This seemingly small point sheds light on why the skin's relationship with diet is so complex. Most of those with psoriasis who also have gluten sensitivity carry the HLA Cw6 gene, which is found only in a subset of people with both conditions.[22] How the skin and the gut interact, and how our skin responds to our diet, partly depends on our genetic make-up. The emerging field of nutrigenetics is revealing that our individual genetic code directly affects our response to nutrients, theoretically leading us towards an age of scientifically personalized diet plans. You are what you eat, depending on who you are.

It is clear, then, that diet does affect skin, but that it's not as straightforward as we'd like it to be. But what about water? Surely the elixir of life is the answer to having a plump, healthy-looking surface? After one of my neighbours returned from a year abroad, I noticed that she was now only ever using her right hand; whether it was walking the dogs, relaxing at the pub or working out in the gym, her left hand always seemed to be fastened to a bottle of mineral water. 'Apparently it's the best way of keeping your skin plump, taut and fresh and your complexion glowing. It even pushes wrinkles out,' she was keen to tell me. On hearing that she had started to drink four litres of water a day, I was surprised she managed to make it out of the bathroom, let alone live her life one-handed. But nowadays it's unusual to see a supermodel without a mineral-water bottle in hand, swearing that their radiant complexion is nourished by a fountainhead of litres of water. It's logical enough, as our skin is made up of cells, which are themselves largely made up of water, requiring regular hydration. However, our other organs need water too, so it's really hard to measure how much of what we're taking in actually gets to the skin, let alone affects its appearance.[23] Despite the ubiquitous internet and magazine recommendations to drink water for healthy skin, there has actually been very little research into this area. This is, perhaps, unsurprising, because water can't be patented so pharmaceutical companies have little to gain from funding such activities. The few existing studies show that a normal daily water intake positively impacts normal skin functions and epidermal hydration.[24] Dehydration makes the skin lose its turgor (its elasticity) and shape because skin cells lose a lot of their volume. We can be certain that a lack of water is bad for our skin, but that doesn't mean that drinking above-average levels is particularly good for it. It's safe to say that drinking recommended daily amounts of water is healthy: roughly 2.5 litres a day for men and 2 litres for women (of which 70–80 per cent comes from drinks and the rest from food). But as this varies according to a person's size, activity level and the environmental temperature, it's not an exact science. Helpfully, our body has a very reliable internal meter: we should drink water whenever we are thirsty.

Water's inebriated cousin, on the other hand, is not so gentle on the skin. When it comes to the appearance and health of our skin, alcohol is almost always detrimental. In the short term it dehydrates the skin, giving it a sallow, puffy appearance, and the sugar that saturates most cocktails also aggravates acne and may even hasten wrinkles. Alcohol's main breakdown product, acetaldehyde, inflames the skin, releasing histamine that dilates blood vessels and causes characteristic facial flushing. Acetaldehyde needs to be broken down by an enzyme called acetaldehyde dehydrogenase, and a deficiency in this enzyme, genetically common in most of those of Chinese, Japanese and Korean ancestry, causes extreme flushing even after one drink, visible in 40 per cent of East Asians.

Long-term effects of alcohol abuse can make their mark on the skin in dramatically visible ways. I remember Terry, a patient on the ward in his early fifties, who was being treated for severe liver cirrhosis. The aged, tired skin on his face seemed dry yet puffy, tinged with the pale yellow of jaundice. Vitamin deficiencies had dried and cracked the skin around his mouth and red spider naevi were littered across his chest. Terry's distended abdomen resembled the skin's crude attempt at painting Caravaggio's *Medusa* – green, engorged veins flowing out from the umbilicus like the serpentine hair of the Greek monster. Indeed, caput medusae ('head of Medusa') is the medical term for this sign of liver failure that blocks up the veins draining the gut into the liver. Small patches of discoid eczema, bubbling up from chronic inflammation, and the distinctive circles of tinea corporis (ringworm), making the most of his damaged immune system, filled in the rest of his abdomen. Any skin that could have been spared had clearly been excoriated and scarred by incessant scratching of the itch caused by liver disease.

Short-term alcohol consumption and the effects of alcohol abuse certainly leave their marks on the skin, but how the skin fares with long-term moderate drinking – say, a glass of wine a night – is less clear. It was once thought that regular alcohol intake contributed to malignant melanoma, the life-threatening skin cancer, but this view didn't take into account what is known in research as a 'confounding factor'. People who have a regular beer are much more likely to practise other risk-taking behaviours, including spending too much time in the sun. When this was accounted for, it was found that there is actually no link between alcohol intake and melanoma, and long-term moderate drinking is most probably safe.[25]

Alcohol is often likened to a truth serum, loosening lips and encouraging guards to be lowered, and it also spills secrets through our skin. When alcohol is metabolized it is partly excreted in our sweat, and this process has been impressively exploited by recent 'transdermal alcohol technology', where a bracelet can now continuously and accurately measure the level of alcohol in our system through the skin. It is not inconceivable that one day we will be adding our blood alcohol content to the wealth of personal data we sell to smart technology.

A sensible, balanced diet helps keep the body and skin healthy (even if there are no wonder-foods), but many people are starting to believe that we can also eat our way to looking more youthful. This is shown by the burgeoning 'nutricosmetics' industry – the supplements and powders whose sales are estimated to reach £5 billion globally by 2020.[26] There has been an explosion in potions promising clear complexions, and alongside drinks containing vitamins and antioxidants there is a new drive to renew elements of the skin from the inside out. If any molecule has the biggest claim to be the elixir to eternal youth, it is collagen. Collagen is a protein that makes up 75 per cent of our skin, giving it structure and plumpness. As we age, levels of collagen in the skin begin to decrease, a process rapidly accelerated by sun damage and smoking. The loss of this skin scaffolding contributes to wrinkles and sagging. It seems logical that treatments replacing this molecule could give back skin's plumpness and flatten out those wrinkles.

Collagen is commonly included in skin creams but the molecule is too big to penetrate our skin from the outside, so any effects are likely to be due to the cream's short-term moisturising qualities rather than the effect of the collagen itself. So can we feed the skin collagen from the inside out instead? In recent years numerous supplements containing collagen in its smaller, hydrolysed form, have claimed to replace collagen, or kick-start its synthesis, in the skin. Many doctors have doubts as to whether it escapes being broken down by our potent stomach acid, and there's just not enough evidence at the moment to know either way. In the study that showed most success, only eighteen women were tested, and as with the chocolate and acne trial, this sample size is far too small to be statistically relevant.[27] In another, larger study, only 15 per cent of the participants had a real improvement in the appearance of wrinkles, something that could be accounted for by other factors.[28] If drinking collagen is affordable and it makes one feel better, then it's not going to do physical harm, though no collagen or antioxidant potion yet has a right to claim to be the elixir of youth. But that's not to say that in the future, with enough evidence, we won't one day be feeding our skin from our gut.

While ingesting collagen is harmless, some nutricosmetics are downright dangerous. In recent years a handful of supplements have reached the market claiming to be a 'drinkable sunscreen'. One claimed to emit 'scalar waves' (whatever they may be), which are supposed to vibrate along the surface of the skin to produce a sun protection factor of thirty. In a subsequent lawsuit, Iowa's Attorney General rightly denounced this quackery as 'almost certainly pure bunk'. Other products seem more plausible, claiming to contain cocktails of antioxidants and vitamins that protect and repair the skin from UV damage. But the evidence for these molecules protecting against skin cancer is limited, and they certainly do not protect against sun damage in the way that sunscreen does. When people lie in the sun believing that they are protected from cancer by a vitamin drink, pseudo-science moves from the realms of comedy to potential tragedy.

While there is little evidence for improving or healing our skin by taking vitamin supplements above recommended daily allowances, the consequences of not having enough can be drastic.

It was the early twentieth century and South Carolina was in a state of emergency. It had started with bright red, scaly rashes appearing on the sun-exposed surfaces of the population's skin. The scales would thicken and darken, and cracks would start to appear. As these red lesions spread, the body seemed to crack open from the inside. Sufferers lay bedridden in excruciating abdominal pain and exhausted from unstoppable diarrhoea. Finally, cracks seemed to open up in their minds. Most would develop depression, headaches and confusion, and for many this progressed to states of psychosis and insanity. Carried away to mental asylums, around 40 per cent of patients with this new, mysterious disease became comatose, then died. Between 1906 and 1914 there were over 30,000 cases in South Carolina alone.[29] The US Surgeon General decided to dispatch Dr Joseph Goldberger, who had made a name for himself uncovering epidemics from Mexico to Manhattan, to get to the root of this deadly new disease. Seeming to appear from nowhere, this epidemic explosion led most in the medical profession to assume that this terrible disease, called pellagra, was an infection. After visiting a number of hospitals, mental asylums and prisons, however, Dr Goldberger noticed an interesting pattern. The only people who seemed to contract the disease were the patients and inmates; none of the staff and doctors ever became infected.

Always one to think outside of the box, Dr Goldberger went about experimenting. In a particularly badly afflicted orphanage – where 172 orphans had the cracked, scaly red skin of pellagra – Dr Goldberger raised funds to introduce a balanced diet of fresh meat, milk and vegetables. All the children were suddenly cured. To further the evidence that diet was the cause, he carried out an experiment in a mental asylum. Over a period of two years, one group of inmates (the control group) continued on their poor diet of cornmeal and wheat, while another group (the intervention group) was given a healthy diet. Both groups were followed up for a period of two years. Half of the control group developed pellagra, compared to none in the healthy-diet group. Despite Goldberger's mounting evidence that diet was the cause of this disease, he was faced with relentless opposition. Many in the medical profession were gripped by the new-found 'germ theory' of disease and still believed pellagra to be an infection. Also, Southern governors and doctors could not tolerate a northerner coming down and blaming the disease on Southern poverty. As a consequence, Goldberger would never live to find out the cause of pellagra, which Conrad Elvehjem found in 1937 to be a deficiency in niacin (vitamin B3),[30] which was available in the meats, vegetables and spices of the asylum group with a balanced diet. As a result, in 1938 niacin was added to bread in the US and pellagra numbers rapidly dropped. It was clear that with pellagra the skin was the harbinger of a dangerously deficient diet.

Goldberger's clinical trials were made possible by another inquisitive, scientific mind researching a different mysterious skin disease a century earlier. In 1740, Commodore George Anson of the British Royal Navy returned from a four-year mission to capture or destabilize Spain's Pacific possessions, from Peru to Panama. Despite the success of the circumnavigation, he returned with a skeleton crew; only 188 of nearly 2,000 men who had set sail survived. The expedition had been ravaged by a different kind of red death: scurvy. No one knew the cause and no one knew the cure. First, reddish-blue spots would appear around the hair follicles, often on the shins. These then slowly enlarged and joined together to form bruises that would eventually cover the whole limb. Cuts and gashes to the skin, a common occurrence on a Navy ship, would take much longer to heal, if they healed at all. These ominous, spreading signs on the skin were accompanied by symptoms of weakness, lethargy and aching legs.

Between the sixteenth and eighteenth centuries, more British sailors died from scurvy than from battle, and a coherent naval approach to treatment was completely lacking. James Lind, a Scotsman born in 1716, joined the Navy as a surgeon's mate and experienced first-hand the devastating effects of the condition. To get to the bottom of this seemingly unstoppable disease, in 1747 Lind performed the first randomized controlled trial documented in medical history. Keeping an open mind, he took twelve sailors and split them into six pairs. Their diet and daily duties were kept as similar as possible, with the exception of a different addition to each pair's diet. Pair number one had cider, pair two had elixir of vitriol (a mixture of alcohol and sulphuric acid), pair three had vinegar, pair four had seawater, pair five had two oranges and one lemon daily, and pair six had a paste wonderfully called 'the bigness of nutmeg'. By day six, only pair five were able to perform active duty and were nursing their other colleagues. James Lind died in 1794, a year before citrus fruit was finally accepted by the Royal Navy as a means to prevent scurvy. His legacy was seen not only in its eradication but also in the founding of the use of clinical trials, which have become a cornerstone of modern medicine. The subsequent daily lime ration for British sailors earned them the name 'limeys' from their American counterparts and helped the British Navy to rule the waves, for a time at least. It wasn't until the early 1930s that Albert Szent-Györgyi in Hungary and Charles Glen King in the USA identified the mystery cure found in citrus fruit: vitamin C.

The skin and the gut talk to each other through diet and metabolism, but they also use another means of communication, which is sometimes dramatic and always intriguing: the immune system. If someone has an allergic response to food, often the first signs show up on the skin, whether it's a red rash, hives or swollen lips. In a genuine food allergy, the body mounts a hyperactive immune response to harmless foods in the form of IgE antibodies. IgE is a type of antibody implicated in many allergies, binding in large numbers to the allergen, travelling to the skin and activating mast cells. Once activated, these mast cells explode, releasing a potent cocktail of histamine and enzymes that lead to the redness and swelling we see on the skin. This means of skin–gut communication is exploited by doctors for food allergy diagnosis. In the skin-prick test, skin is pierced with a lancet to let in a tiny amount of the food allergen. A small area of itching, redness and swelling suggests an underlying allergy to that particular food. Food allergies also influence skin disease. If someone has a clinically proven food allergy, there is some evidence that an exclusion diet not only reduces the chances of an allergic reaction but also alleviates symptoms of skin conditions such as eczema.[31]

Pseudoallergy, a skin reaction or intolerance resembling allergy but where IgE is not produced, is harder to diagnose. In chronic urticaria (commonly known as hives), one theory is that pseudo-allergens in the diet – found in additives and preservatives as well as natural compounds such as salicylic acid, found in plants – directly react with the skin without an antibody reaction. Although this theory is controversial, pseudoallergen-free diets have been shown to work in a subset of chronic urticaria patients, even when other treatments have failed.[32, 33]

I once saw a young woman who, alongside unexpectedly losing weight over the previous few months, had broken out in 'insufferably itchy' tiny blisters that had erupted symmetrically over her buttocks and the backs of her arms and legs. This was dermatitis herpetiformis (resembling a herpes rash), a bubbling eruption on the skin emanating from coeliac disease in the gut. While in an allergy an IgE antibody responds to common food allergens such as milk, eggs and shellfish, in coeliac disease a form of antibody called IgA – which protects the gut and mucous membranes from foreign invaders – reacts against gliadin (amongst other molecules), a protein found in gluten. In coeliac disease, the immune system also attacks a molecule in the gut called tissue transglutaminase, and it is likely that it subsequently recognizes a very similar molecule in the skin (epidermal transglutaminase) and starts to form antibodies against it. IgA antibodies then travel from the gut to the skin and are deposited in the top section of the dermis, causing the itchy, blistering, inflamed skin.

In coeliac disease, the immune bridge between the skin and the gut is fairly well understood. In many other diseases, such as the alarming-looking skin nodules and ulcers associated with inflammatory bowel disease (including Crohn's disease), we know very little about which roads link – and which molecules and immune cells travel between – these two distant organs. Rosacea, the inflammatory skin disease often involving *Demodex* mites and characterized by facial flushing, nodules and swelling, is also associated with a number of gastrointestinal diseases. Among these, rosacea seems to be caused by an excessive growth of bacterial populations in the small intestine, a condition known as small-intestinal bacterial overgrowth (SIBO). If SIBO is treated with an antibiotic that affects only the gut, rosacea on the skin is cleared.[34]

SIBO demonstrates another mysterious link between the skin and the gut: namely, how the gut microbiome influences the skin. There are more bacteria living inside us than we have human cells; this complex civilization of living microbes in our gut is sometimes called 'the forgotten organ'. New research is slowly revealing how the composition of our bacterial flora influences our health, and this is certainly only the tip of a microbial iceberg. Changing the course of skin disease and avoiding the use of antibiotics by adjusting gut bacterial populations is certainly an attractive prospect.

Élie Metchnikoff, who was awarded the Nobel Prize more than a century ago, was a man ahead of his time. The Russian zoologist believed that 'eventually, it may become possible to restore the health of a depleted microbiome simply by swallowing a capsule crammed with billions of bacterial cells, or by eating yogurt', predicting the use of probiotics decades before it became fashionable or taken seriously by science. More recently, children with eczema have been shown, on average, to have a reduced gut-microbe diversity and low levels of good gut bacteria, including *Lactobacillus*, *Bifidobacterium* and *Bacteroides*, and recent large-scale studies suggest that probiotics containing a mixture of *Lactobacilli* and *Bifidobacteria* do improve eczema symptoms in children.[35] It is also becoming evident that if a mother takes probiotics during pregnancy, it reduces the risk of her child developing atopic eczema between the ages of two and seven years.[36]

'Prebiotics' (not to be confused with probiotics) do not contain bacteria but instead consist of non-digestible elements that support the growth of 'good' bacteria and modify the environment of the gut. For instance, dietary fibre, found in fruits, vegetables and grains, is a great prebiotic food source. 'Synbiotics' is the name given to a combination of probiotics and prebiotics, and early results for the treatment of skin diseases with synbiotics are looking promising. A 2016 study found that eight weeks of treatment with an oral symbiotic reduced eczema symptoms in children aged one and above.[37]

The ripples made by the billions of bacteria in the gut wash over on the shores of the skin, but via a number of different routes. The first is by altering the immune system. In a study in which mice drank the probiotic *Lactobacillus reuteri*, natural anti-inflammatory molecules in the skin were increased.[38] Equally, an aberrant gut microbiome can negatively influence the skin's immune system. It is becoming clear that an imbalance of microbes in the gut, known as gut dysbiosis, causes a more permeable gut lining that enables pathogenic and inflammatory molecules to enter the bloodstream and damage distant sites of the body, known as the 'leaky gut hypothesis'.[39] There is mounting evidence that skewing the immune system through dysbiosis causes inflammation in the skin–gut–joint relationship, worsening psoriatic arthritis, an inflammatory arthritis that occurs in those affected with psoriasis.[40] In psoriasis patients, dysbiosis raises bacterial DNA and inflammatory proteins in the blood, and newborn mice suffer worsened psoriasis when given antibiotics to reduce the diversity of their gut microbiome.[41] In 2018 a team of French researchers also found that negatively altering the gut microbiomes of mice increased the frequency and severity of allergic reactions on the skin.[42]

The second means of communication between the bacterial populations on these distant organs is diet. Human gut microbes are essential in the breaking-down and metabolism of food. They ferment dietary fibre into short-chain fatty acids, which are then absorbed by the body and exhibit anti-inflammatory properties. Some believe that the skin could even be a reservoir for fats synthesized in our gut by the microbiome. It could also be possible that diet affects the make-up of the microbes on our skin. As I discovered in the curry house, garlic metabolites (such as allyl methyl sulphide) are excreted through the skin and are known to have antimicrobial properties. Finding out whether specific components of diet affect our skin is difficult, however, and the fact that each individual's gut microbiome is subtly different adds another layer of complexity.

But the intricate channels back and forth between the skin and the gut may not stop there. At medical school, when one of my housemates went through a period of mental or emotional stress he would simultaneously break out in a rash similar to eczema and lie on the sofa doubled over in aching, agonizing irritable bowel syndrome. In my quest to have a disease named after me (and 'Lyman Syndrome' certainly has a nasty ring to it), I hypothesized that all three organs – skin, brain and gut – were undergoing mutual distress. Unfortunately the dermatologists John H. Stokes and Donald M. Pillsbury had managed to get there eighty years before me, outlining an 'important linkage of emotion with cutaneous outbreaks of erythema, urticarial and dermatitis by way of the physiology of the gastrointestinal tract'.[43] No one would argue with the Roman poet Juvenal's assertion about the benefits of 'mens sana in corpore sano' (a healthy mind in a healthy body). Our mental state certainly affects our skin as well as our gut (as we will see in Chapter 7), but it is becoming increasingly evident that inflammation in the gut contributes to inflammation in the brain, influencing one's mental state and exacerbating anxiety and depression.[44] It is currently untested, but plausible, that this gut-altered mental state can even affect our skin. It also works the other way: mental stress alters the make-up of the gut microbiome. One study showed reduced populations of 'good bacteria' *Lactobacillus* and *Bifidobacteria* in response to mental stress in animal models.[45] Certain gut bacteria can produce neurotransmitters: *Streptococcus* and *Candida* produce serotonin, which stimulates gut contraction, and *Bacillus* and *Escherichia* produce noradrenaline, known to dampen down digestive activity. Psychological stress reduces gut contractions and mobility, potentially leading to bacterial overgrowth and increased gut permeability, which can go on to influence skin disease such as rosacea.[46, 47] This brain–gut–skin axis is a little-explored frontier that represents one of the intricate and mysterious ways in which our mind, body and microbial companions communicate.

It is apparent, then, that what we eat affects our skin from the inside out, but a recent and radical piece in the skin–gut jigsaw reveals that it also works the other way: skin 'eats'. New evidence has shown that what lands on our skin as a child may directly lead to food allergies. In 2015 the first results of the landmark LEAP (Learning Early About Peanut allergy) trial were released. The LEAP study demonstrated that for infants at high risk of developing peanut allergy, such as those with an egg allergy or eczema, eating snacks containing peanuts prevented the development of a peanut allergy later in childhood.[48] This strongly supports the idea of oral tolerance, where the body learns to respond protectively to harmless peanut molecules instead of mounting a vigorous allergic response to them. We now understand that tolerance can be achieved through our gut – but what happens if food particles are 'eaten' by our skin first? This isn't as far-fetched as it sounds. When our skin barrier is compromised, most commonly in eczema, it is easier for airborne food particles that land on the skin to sneak past this barrier. These tiny allergens are then 'eaten' by immune cells called phagocytes that reside in the skin, in a process called phagocytosis. The phagocytes then communicate with other immune cells to cause 'sensitization', in which the immune system recognizes these food particles as foreign and the body becomes primed against these foods.[49] Any further exposure to the food results in an allergic reaction. Infants with eczema are much more likely to develop subsequent food allergies – to peanuts, egg or cow's milk, for example – than those with healthy skin. In fact, eczema is the greatest risk factor for food allergy in infants, and the skin condition usually precedes food allergy.[50] A 2018 study strengthened these findings by showing that food allergy is triggered by a combination of three skin-eating factors: genetic influences that increase skin absorbency; exposure to dust and food allergens in the household; and an overuse of infant wipes that leave soap on the skin, disrupting the lipid barrier.[51] It's not yet known to what extent (if any) wet wipes have contributed to the rise in childhood food allergies in recent years, but evidence suggests that their overuse is detrimental to an infant's skin barrier. As research builds, it seems possible that treating eczema early on and making changes to protect a baby's skin barrier could potentially prevent food allergies later on in life.

The mysterious displays on the skin that bubble up from the gut, whether an immediate reaction to a food allergen, the nodules and ulcers that accompany inflammatory bowel disease, or the attractive glow of a vegetable-rich diet, show that these distant organs really do communicate with each other, and science is slowly finding out how. Unfortunately, when it comes to feeding our skin, there are few silver bullets, but the good news is that the answer to skin health is the answer to general health: a sustainable, balanced diet. The skin is a herald of health, and it calls us to respect the intricacy and beauty of our body.

4

Towards the Light

The story of our skin and the sun

'Let me warn you, Icarus, to take the middle way, in case the moisture weighs down your wings, if you fly too low, or if you go too high, the sun scorches them. Travel between the extremes.'

OVID, *METAMORPHOSES*, BOOK VIII

AN OVERAMBITIOUS MORNING jog along the shores of the Greek island of Samos was taking its toll. Midway through an attempt to reach the top of the headland, I sat down to catch my breath and survey the bay below. The dawn light was beginning to crawl across the Icarian Sea, the fabled resting place of the boy who flew too close to the sun. Down on the beach, sand was barely visible between the tightly ordered rows of sun loungers – pews and prayer mats ready for thousands of sun-worshippers.

This language of divinity around the sun isn't new; our closest star has been worshipped for thousands of years. It brings life, light and healing powers, but it also demands reverence. A direct look with the naked eye can cause blindness, and long exposure burns the skin and exhausts the body. Modern science has only increased our awe and respect for this object that is 330,000 times heavier than the Earth and takes up almost 99.9 per cent of the mass of our solar system. The Ancient Greeks made Apollo god of the sun, but rather tellingly he was also god of both healing and of disease. We all have an instinctive feeling that the sun, this powerful two-faced deity, can be both good and bad for us. So should we revere it or fear it?

You've taken a well-needed holiday to a similarly sunny island and have slept on the beach for a good part of the afternoon. On returning to your hotel, you notice in the mirror that the tip of your nose has slightly reddened; you have been sun-kissed. To find out what has happened in physiological terms, we need to track the particles of sunlight responsible, known as photons, on their commute to work. Let's skip over the journey taken by a photon from the centre of the sun to its surface, as the slow process of nudging forward amongst countless other photons can take an estimated 100,000 years. Once released from its history, if this tiny particle happens to escape from the surface of the sun in the direction of Earth, it has a relatively short commute. Travelling at 671 million mph, sunlight takes eight minutes and seventeen seconds to reach the Earth. If you're sitting outside reading this book by virtue of natural light, the energy illuminating the page probably left the sun two to four pages back. While the vast majority of photons striking our skin are emitted from the sun, a tiny proportion originate in other stars. A team from the University of Western Australia even calculated that about ten trillionths of a suntan comes from stars found in other galaxies, although admittedly you'd have to be basking in this light for a few trillion years to notice its effects.[1]

Importantly, in the same way that crowded commutes bring together a diverse population, the sunlight that hits your nose is made up of different photons, each with differing characteristics determined by its wavelength. The human eye can detect the wavelengths of visible light, enabling you to see the slight sunburn on the tip of your nose. The particles of sunlight that actually bring about the burn, however, are invisible, high-energy, ultraviolet (UV) rays. The majority of UV light that reaches the Earth's surface is made up of UVA particles. These penetrate beyond the outer layer of our skin and cause damage to the deeper dermis. Over time this weakens the skin's supporting layer of collagen and elastin, causing wrinkles, leathery skin and pigmented spots in a process known as photoageing. Although it contributes to tanning, UVA is not responsible for sunburn and was originally thought not to cause cancer. This is why it has traditionally been used in sunbeds. But evidence is beginning to show that UVA can initiate and accelerate skin cancer development, as well as speed up the ageing process.

The most notorious character that commutes along a ray of sunlight, however, is UVB. This double-edged solar sword delivers both the sun's pain and provision. As a high-energy particle, UVB hits the outer epidermis of the skin and slices DNA apart. The immediate response from the skin is inflammation, seen in redness, swelling and blistering. As well as chopping up DNA, UVB also splits apart vitamin D precursors in the skin. Precursors are inactive forms of a compound that, when broken up in a specific way, release the active substance. As a result, UVB radiation is one of our most important sources of this essential molecule. (The most powerful and dangerous rays of sunlight, UVC, would devastate our skin, but we can thank the Earth's own skin, our atmosphere – made of ozone and oxygen – for taking the hit.)

By slicing through DNA, these waves of UVB radiation should leave us all with unremitting skin cancer and early graves. So what stands in their path? The answer is the humble melanocyte. These are small, octopus-like cells that dwell at the bottom of the epidermis. Like octopuses, they spew out ink, called melanin. These black, brown and red pigments consist of different types of complex polymers, a diversity that means they can absorb almost any wavelength of ultraviolet light. This remarkable pigment is our natural sunscreen, and a 2014 study at Lund University in Sweden discovered the remarkable way in which it works.[2] When a melanin molecule is struck by UV light, it fires off a proton, disarming the UV ray's power and transferring it into harmless heat – all in one thousandth of a billionth of a second. A bombardment of sunlight kicks our melanocytes into action, and over a period of two to three days after exposure we form a tan as a protective response to

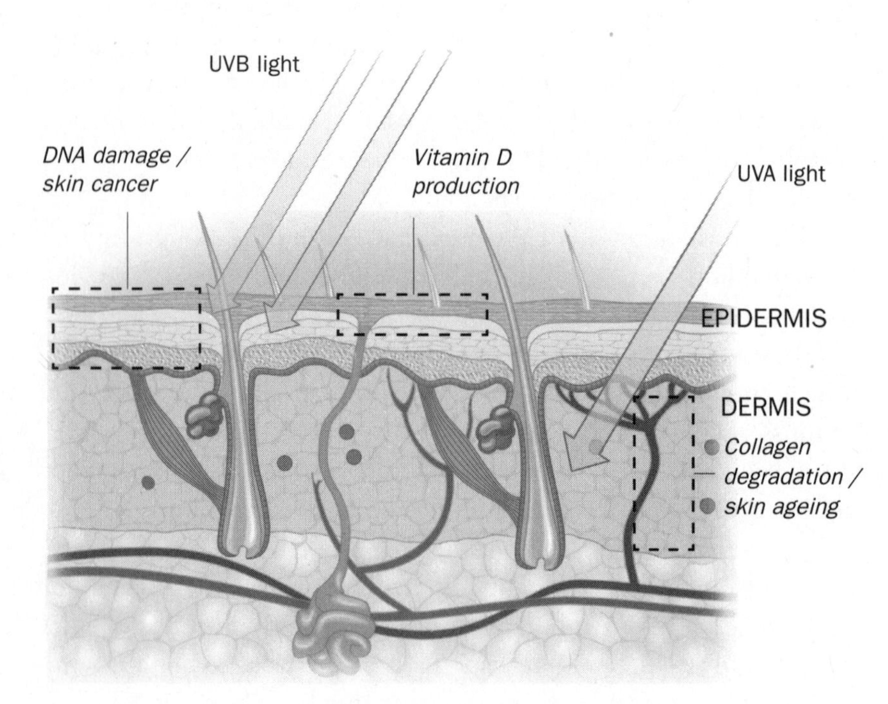

UVA AND UBV LIGHT

damage – a callus of colour, if you like. Our long-term, background skin pigmentation, however, is largely determined by the type and concentration of melanin in the skin, creating humankind's sepia rainbow. Dark skin does not have a larger number of melanocytes than light skin; instead, the skin's octopuses work harder at producing more of the protective pigment. To fully understand melanin's vital importance is to see the lives of those who have a complete lack of it. People with albinism – who have a genetic absence of melanin – are subject to all forms of skin cancer from an early age, and without lifelong sun protection as well as meticulous attention and treatment, their lives are cut tragically short.

Even though melanin is the original sunscreen, a suntan is not a sensible form of sun protection: it provides a sun protection factor (SPF) of around only 3 and leaves a trail of DNA destruction in its wake. So, contrary to popular belief, a pre-holiday top-up on a sunbed does almost nothing to protect the skin from the sun. Sun exposure damages both our health and our looks; it is actually a more significant cause of accelerated skin ageing than all other causes combined. Most importantly, however, it is the greatest risk factor for a group of diseases that is causing suffering and death at epidemic levels in many parts of the world: skin cancer.

My first encounter with skin cancer is etched into my memory. I sat in the corner of the consultation room as an oncologist assessed Catriona, a thirty-year-old Irish woman with stage IV malignant melanoma. Gaunt and emaciated after months of cancer and chemotherapy, she looked decades older than the fit, young teacher who until recently competed in triathlons every weekend. A year earlier she had noticed a small, flat area of red, brown and black pigmentation on the skin above her right shoulder blade. It started to grow, and even despite a relatively early diagnosis and every attempt at treatment, the melanoma spread to her lungs, liver and bones. When I saw her she had just been given between six and twelve months to live.

'He left me alone – he left me to die!' she sobbed that day. In fits and starts, Catriona explained to the oncologist what she thought had caused her cancer. Twenty-two years earlier, when she was eight years old, her family had gone on a summer holiday to the Spanish coast. 'We wore very little sun cream,' she explained. 'Mum wanted to look beautiful and Dad said that a tan was healthy. Anyway, on the first afternoon, Mum went shopping and Dad and I went to the beach. Dad went off to the pub and left me on my own for – I don't know – maybe four or five hours. When we eventually went back to the hotel in the evening, the sunburn here was so bad – it was blistering and bleeding – that I had to go to the doctors.' She was pointing to the place where she had first spotted the emerging melanoma two decades later.

Could one bad sunburn as a child have caused Catriona's terminal cancer? Current research suggests that although many people who are severely sunburned as children never develop skin cancer, one blistering sunburn in childhood increases the risk of melanoma later on in life by 50 per cent.[3] Another study suggests that white women (the study did not look at men) who get five or more severe sunburns in their teens have double the risk of developing melanoma.[4] Once the DNA damage is done, the area of skin is particularly vulnerable throughout life. Unrepaired DNA damage caused by sunlight leads to mutations in the DNA sequence, which could develop into a life-threatening cancer. The significance of the sun's role in skin cancer, particularly in countries with large populations of people with light skin tones, is wildly underestimated.

After Catriona had left, the oncologist turned to me. 'As humans we're notoriously bad at assessing risk, particularly when it comes to our future health,' he observed. 'But when it comes to someone under your care . . . I think that a bad sunburn on an infant is nothing short of child abuse.'

Over the past thirty years, more Americans have had skin cancer than all other cancers combined,[5] and as many as two in three Australians will develop skin cancer over the course of their lives.[6] Skin cancer particularly affects light-skinned North Americans and Europeans driven by the desire to have a 'healthy tan' and aided by the rise of cheap holidays in the sun. Particularly in the Western world, the explosion in skin cancer rates over the past few decades is nothing short of a public health crisis, with increased death and suffering as well as mounting healthcare costs.[7] Treatment for skin cancer alone is predicted to cost the National Health Service £500 million a year by 2025; meanwhile, the increase in skin cancer cases is testing the capacity of dermatology departments, with waiting lists lengthening for all skin diseases.

There are three main types of skin cancer. The first is the basal cell carcinoma (BCC), which is by far the most common. These pearly lumps, often found on the sun-exposed areas of the head and neck, very rarely cause death, but if they are not removed they can damage nearby structures such as eyes and ears. Then there are the squamous cell carcinomas (SCC), which can appear crusty and ulcerated and often bleed. SCCs are less common but more dangerous, with a greater chance of metastasizing (spreading to other parts of the body) compared to BCCs. The most concerning skin cancer, however – and the one with the highest mortality – is melanoma. Although less common than the other two forms of skin cancer, the incidence of this deadly cancer is increasing. In the US in 2018, there were roughly 90,000 new melanoma diagnoses, with almost 10,000 deaths – a fifteen-fold increase in the past forty years.[8] In the UK, the incidence of melanoma is predicted to increase by 7 per cent by 2035.[9]

An early diagnosis of melanoma is particularly important. In the 1970s, five in ten people with melanoma died, but this number has dropped to one in ten, largely due to a greater public awareness for spotting skin cancers and earlier diagnoses. A 2017 study from the Cleveland Clinic emphasizes that early detection of melanoma is critical for survival outcomes.[10] As well as skin type, 'moley-ness' also influences melanoma risk. More than eleven moles on an individual's right arm suggests more than one hundred exist on the whole body, increasing the risk of melanoma, as 20 to 40 per cent of melanomas form within existing moles.[11] 'Atypical' moles can also run in families, where their presence is termed 'dysplasic naevus syndrome'. Moles present at birth, known as congenital melanocytic naevi, have up to a 10 per cent chance of developing into melanomas later in life. A relatively simple way for the layperson to detect melanomas lurking amongst harmless moles is by looking out for the 'ABCDE' features:

- Asymmetry
- Irregular Borders
- More than one Colour within the patch
- A Diameter of more than 6mm
- Its Evolution (the way a mole changes in colour or size)

I also like to think of E as standing for 'ask an Expert'. General practitioners and dermatologists have the experience and training, not to mention being able to assess the lesion under a dermatoscope – a specialized magnifying glass for examining the skin – to determine whether a mole might be malignant or not. This is supported by a 2018 study led by the University of Birmingham, which looked at the worldwide body of research on the topic and found that visual inspection alone was not a satisfactory way of detecting melanomas, with an expert trained in dermatoscope use missing far fewer lesions.[12] It also found that smartphone applications commonly used by the public to assess suspicious moles had a high probability of missing melanomas.

The colours, freckles and spots of our skin tell a story, and it's one that we should read regularly. One way of attempting to understand the skin's risk of developing skin cancer is categorizing the skin into 'phototypes', as shown in the Fitzpatrick scale:

- Type 1: Pale white skin. Always burns, does not tan.
- Type 2: Fair skin, darker eyes. Burns easily, tans poorly.
- Type 3: Darker white skin. Tans after burning.
- Type 4: Light brown skin. Burns slightly, tans easily.
- Type 5: Brown skin. Rarely burns, tans darkly.
- Type 6: Dark brown/black skin. Never burns, always tans.[13]

The Fitzpatrick scale is flawed in that it so crudely categorizes the countless hues of the sepia spectrum of human skin colour; and in that dark brown and black skin certainly can burn, although it is much less likely and less severe than with lighter skin. The scale is a good guide for risk, however, and has also played an unexpected role in promoting fair racial representation: the scale formed the basis for the addition of five skin tones to emojis, the modern hieroglyphics used in electronic communication.[14]

While skin cancer is the most common form of cancer among the Caucasian population, skin cancer affects all skin tones. UV light can damage any skin colour, and there are other risk factors leading to skin cancer, from genetics to smoking. Studies in the US show that although malignant melanoma is much less common in people with black skin than in the white population, black people have a considerably worse survival rate than whites. The reason for this imbalance is not completely known, but it could well be a combination of reduced access to healthcare amongst the black population in the USA and a lack of awareness in both the medical profession and black communities that cancer in black skin is even a possibility. A famous case of misdiagnosis of melanoma is that of reggae singer Bob Marley, whose fatal melanoma on his toe is said to have been initially misdiagnosed as a soccer injury.

Reading about these unexamined inequalities in outcomes of skin cancer sufferers, I decided to test whether there was a difference between doctors' diagnoses of melanoma on white and black skin types in primary-care settings in the UK. Sourcing patients from the NHS, a health service free at the point of delivery, enabled a more level playing field in socio-economic terms than the system on offer in the US. In collaboration with two universities in the UK, I devised a twenty-photo picture test entitled 'Primary Care Dermatology Quiz' and sent it via email to 3,000 general practitioners across England.[15] A range of different skin diseases were presented, together with a drop-down list of twenty diseases from which GPs could choose their answers. What I did not tell the participants was that I was only interested in four particular photos, which I placed randomly within the quiz. Two were of melanoma on white skin and two were of melanoma on black skin. Interestingly, the doctors correctly identified melanoma on white skin around 90 per cent of the time, but only just over 50 per cent of the time on black skin. This small study has a number of limitations, but it offers a glimpse into the need for medical education on how this potentially deadly skin disease differs across the scale of human skin colour.

Skin cancer is clearly a growing problem, so what can we do to reduce our chances of developing it? The first thing to appreciate is that the 'healthy tan' is a myth. There is overwhelming evidence that skin damage, even from mild tans, accumulates over the years. The ancient Egyptians, although they worshipped the sun, realized its dangers and developed the earliest recorded form of sunscreen. Their recipe of rice bran and jasmine does in fact contain molecules known to heal damaged skin, but thankfully in recent years we have developed much more effective sunblock preparations. To have adequate protection while out enjoying, or working in, the sun requires all exposed skin of an average-sized adult to be covered with 35–45ml (the size of a golf ball or 6–8 teaspoons) of broad-spectrum (UVA and UVB) sunscreen with an SPF of at least 15 – meaning it will take fifteen times longer for the skin to redden than it would without sunscreen. In the UK, studies show that there is enormous confusion over sunscreen labelling. SPF accounts only for UVB radiation; a separate 1–5 star system indicates a sunscreen's protection against UVA. Nearly half of those surveyed didn't know what SPF means and only 8 per cent knew that it relates to UVB radiation only.[16]

Other common-sense tips for reducing the risk of skin cancer include avoiding sunbeds, seeking shade, wearing hats and appropriate clothing, and teaching children sun-cream application.

One country has shown that when sun protection becomes second nature it really does work as a preventative measure against skin cancer. Many Australians are of British descent, their pasty ancestors transplanted from their dark, rainy home on the shores of Northern Europe to a hot, sun-kissed continent at the other side of the world. Unsurprisingly, then, Australia is the skin cancer capital of the world, but over the past thirty years it has also managed to be the only country to reduce skin cancer rates. A colleague of mine left Australia for the UK in 1980. After returning from visiting family in Sydney in 1985 he remarked, 'Not much had changed down under, although in the seventies all my mates had long hair and walked topless around town. Now they all had short hair, wore T-shirts and hats and were slathered in sunscreen. Sid the Seagull really is part of the Australian psyche now.' Sid was introduced to the television screens of Australia in 1981, singing the frustratingly catchy jingle: 'Slip, slop, slap! Slip on a shirt, slop on sunscreen and slap on a hat!' The Slip-Slop-Slap campaign is one of the most successful public health awareness drives in history and has long been admired by marketing companies and healthcare organizations worldwide.[17] It shows how a well-delivered, uncomplicated message can raise awareness of an issue and ultimately convert that knowledge into action.

Nevertheless, the gap between public knowledge of sun damage and the actual use of sun protection – even in Australia – shows that it takes a lot to change our attitudes to health. A large survey conducted by the British Association of Dermatologists in 2015 found that, in the UK, 80 per cent of the population is worried about skin cancer, yet 72 per cent have been sunburned in the past year.[18] Similarly, a 2017 study looking into the sun-protection behaviours of 20,000 individuals from twenty-three countries around the world found that although nine in ten people are aware of the link between sun exposure and skin cancer, almost half of respondents don't take any measures to protect their skin on holiday.[19] Psychological studies into sun-care campaigns intriguingly show that appealing to our vanity is actually more effective than addressing our health. When people are shown photos of skin cancer and are told that sun damage will affect their health in the future, it does not seem to affect their behaviour. However, if people are shown photos of wrinkles and freckles caused by sun damage and are told that tanning negatively affects their future appearance, they are much more likely to adhere to sun protection guidance.

Why does society – in particular white, Western society – seek the sun? In pre-1920s Europe and America, tanned skin was associated with the lower classes, who had to toil in the fields, whereas pale features were deemed attractive. This perception of societal beauty still exists in many developing nations, particularly in areas of Africa and Asia (as we will explore further in Chapter 9). But ever since Coco Chanel accidentally caught too much sun in the French Riviera, and soon afterwards declared in *Vogue* magazine that 'The 1929 girl must be tanned', there has been enormous societal pressure for young Westerners to have a 'healthy glow'. Tans became a badge of holiday leisure and wealth, rather than toil. This cultural shift in how we wear beauty and status on our skin is the driving force behind the rising skin cancer statistics.

Alongside societal pressure, sunlight can literally be addictive. Like a drug, sunlight has both positive and negative effects on the body and, like a drug, it can also get us hooked. 'Tanorexia' is a real phenomenon, where sunlight exposure induces the skin's synthesis of β-endorphin, which enters the bloodstream and causes an opioid-like effect; opioids being the pain-relieving and addictive family to which morphine and heroin belong. In fact, 20 per cent of beachgoers show signs of sun dependence that would satisfy the symptom criteria for addiction and substance abuse.[20]

In an ideal world, people would not want to change their skin colour to appear more attractive. But for those who want a 'healthy tan' without the UV-induced side effects of premature ageing and skin cancer, the big question is whether there is a glowing El Dorado where one can achieve a natural-looking tan without the effort and orange hue of one-day bronzers. An unexpected solution is one that can be eaten. A diet rich in colourful, carotenoid-containing vegetables such as carrots and tomatoes can marginally (but noticeably) deliver a golden glow, and the perceived attractiveness of one's complexion after a carotenoid-rich diet is higher than for someone with a mild suntan.[21] Interestingly, participants who are told that eating fruit and vegetables brings about a desirable skin hue are more likely to stick to this healthy diet than if they are told that it will reduce heart attacks in the future. Perversely, again it seems easier for us to prioritize present appearance over potentially life-threatening consequences in the future.

In 2017, however, a game-changing breakthrough was made

that could genuinely deliver a 'real fake tan'.[22] A small molecule called a SIK inhibitor has been shown to switch on melanin production in melanocytes, naturally increasing the amount of this protective pigment. There is still a long way to go but, if successful, this could enable a sunless, protective tan, which would be particularly helpful for those with the fairest, 'Type 1', skin.

For the vast majority of us, it takes a lot of DNA damage and repeated exposure to the sun's rays to cause skin cancer. This is partly because we have an elegant process for healing the DNA damage caused by UV light. In 'nucleotide excision repair', a protein complex travels along the strands of DNA and, like a thorough editor, spellchecks it for mistakes in the code, specifically those caused by UV damage. If there is an error, the protein complex binds to the intact strand of DNA opposite the damaged area. It then recruits scissor-like proteins, which make incisions in the DNA upstream and downstream of the damaged strand, which subsequently falls away. The correct code is then rebuilt and glued back on to the strand of DNA by the enzymes DNA polymerase and DNA ligase respectively. It seems that these intricate processes mainly occur in the evening, governed by our internal body clock, to reduce mutations caused by UV rays during the day.

Sadly, like many discoveries in medicine, we know a lot about this process from the few people who are unfortunate enough not to have it. During a visit to an African specialist skin hospital in 2016, I was introduced to a ten-year-old girl and her six-year-old brother. The girl's face was pocked with lumps, freckles and scars from previous surgeries. A patch covered her left eye, which had been lost a month before due to an invasive basal cell carcinoma. Her younger brother's face was stained with patches of pigmentation and the emergence of odd, suspicious-looking lumps. Neither had the cheekiness or vitality usual in children of their age; they were completely subdued by their condition. The siblings had xeroderma pigmentosum, a genetic condition in which the intricate DNA-repair mechanism responding to UV light is completely absent.[23] In sufferers' short lives (in the developing world, very few with this condition live beyond their teens) the development of skin cancer is horrifically accelerated. Almost any exposure to sunlight results in immediate sunburn.[24] The Tanzanian doctor showing me around the hospital called them 'the moonlight children'; while in many parts of the world, society banishes those with visible skin diseases to the dark, for sufferers of xeroderma pigmentosum, staying in the dark is the only treatment. In Europe and the USA, roughly one in a million people have this devastating disease, but at least there are the resources to help them. At Camp Sundown in upstate New York, the clocks are turned upside down so children with xeroderma pigmentosum are able to play outside and engage in activities under the cover of darkness. Sitting beside the two cancer-ridden children in Tanzania, however, I could not begin to imagine a life in a hot, developing African country, on the run from every ray of sun.

One of the highest prevalences of xeroderma pigmentosum in the United States lies in an arid corner of Arizona. The Native Americans of the Navajo Nation are thirty-three times more likely to have the disease compared to the rest of the country. Many Navajo medicine men believe that this brutal, visible disease is a curse from their ancestors. Some geneticists believe that, in one sense, there may be some truth in this. The 1860s saw tensions between US government forces and the Navajo tribes reach tipping point, resulting in a series of battles that culminated in the US government forcing the whole Navajo population to walk, at gunpoint, from their homelands in Arizona to Bosque Redondo in New Mexico, some three hundred miles away – an event known as The Long Walk of the Navajo. In the battles, disease and famine of these years, the Navajo population of reproductive age dropped from roughly 20,000 to just 2,000. This rapid reduction caused a genetic bottleneck, meaning that most of the

250,000-strong Navajo population today is descended from this tiny pool of recent ancestors. Among these 2,000 ancestors there happened to be an unusually high percentage of people carrying the faulty xeroderma pigmentosum gene.[25] Not only does the memory of The Long Walk of the Navajo play a defining role in the identity of these people, but the scars of history also act as a constant physical reminder. Our genetic code is an archive of detailed historical information and it writes stories on our skin.

It is not only genes that help or hinder the effects of sunlight. When I was fourteen, my return to school after the summer holidays was marked by the absence of my friend James. When he eventually returned after what seemed an age (though it was probably only a week), he was a completely different person. Enveloped in a long-sleeved T-shirt, he would no longer come out to play football in the afternoon. In fact, he never left the school building. It took another few weeks before he revealed the reason for his newfound aversion to the outdoors. Two days before the start of the new school year, James had been 'forced' to carry out gardening chores at his parents' house. It was a rare hot English summer's day. He emerged into the sunlight after clearing away overgrown bushes in a small shaded copse and – in his words – 'immediately began to melt'. Severe, deep, inexplicable blisters and burns appeared across his exposed arms and the back of his neck.

All was explained after treatment at hospital, where the culprit was found to be the giant hogweed, a close relative of cow parsley. It is originally from southern Russia and Georgia, but thanks to the British love for ornamental plants it is now an invasive weed marching steadily across much of Europe and North America. It looks harmless and dull, but when its sap combines with the sun's UV rays it induces 'phytophotodermatitis': phyto (plant), photo (light), dermatitis (skin inflammation). Giant hogweed contains molecules called furanocoumarins that are found in certain plants and fruit, and are phototoxic: poisonous when exposed to UV rays. The resulting inflammation resembles a painful chemical burn that can have serious effects on the skin, such as long-lasting scars and discolouration.

After leaving school, the same luckless friend worked behind a bar on the Spanish island of Mallorca. Following his usual day shift, he was woken up a few hours into his sleep by a searing pain across his right hand: it was littered with bulging blisters and red, bleeding skin. He was soon to find out that this second-degree burn was a case of 'margarita dermatitis'. While he was making the popular eponymous cocktail for guests by the poolside, the sunshine and lime juice had made a potent cocktail of their own. The phototoxic molecules in limes (which are also found in lemons) reacted with UV light, causing a similar reaction to the one he had been subjected to six years earlier.

The sun harms, but it also heals. Outside the prestigious Rigshospitalet hospital in Copenhagen stands an unusual, imposing monument. Bronze statues of three naked figures bestride a granite rock: a standing man flanked by two kneeling women, their bodies contorted and strained upwards like flowers growing towards the sun. *Mod Lyset* ('Towards the Light') was made by Rudolph Tegner in 1909 to celebrate the work of the Faroese-Danish doctor Niels Finsen, the father of modern phototherapy.

The statues, in their heroic Ancient Greek poses, embody the newly discovered healing powers of light, but also nod to long-forgotten knowledge. Vitiligo is an intensely visible skin disease characterized by localized patches of skin completely lacking in pigment. A definitive cause is not yet known, but it is most likely brought about by interplay between genetics and the destruction of melanocytes by our own immune cells.[26] The three-and-a-half-thousand-year-old Egyptian Ebers Papyrus describes the use of *Ammi majus*, a Nile Valley plant, in a powder that, when applied to the white, unpigmented skin of vitiligo with the skin exposed to the midday sun, permanently brought back pigmentation. Similar ancient texts from India and China confirm beliefs that combinations of plants and sunlight could cure skin disease. The Ancient Greek 'Father of Medicine', Hippocrates, who was almost certainly influenced by his visits to Egypt, was also swayed by the healing powers of the sun. It is not surprising that the Ancient Greek, Roman and Celtic sun gods are also closely associated with medicine and healing. The word heliotherapy (now commonly called phototherapy) derives from the Ancient Greek sun god Helios, whose job it was to drive the sun across the sky each day in what must have been a sizeable chariot.

We have to wait almost 2,000 years before we find Dr Niels Finsen, the Danish light pioneer, positioning his patients in his newly built 'sun garden' in Copenhagen at the start of the twentieth century. Firmly believing in the healing powers of sunlight, Dr Finsen was testing its effect on lupus vulgaris, a painful and disfiguring skin infection caused by the bacterium *Mycobacterium tuberculosis*, when his inquisitive mind led him to investigate whether specific wavelengths of light delivered distinct healing properties. He began to find that UV light was curing many of his patients of their lupus vulgaris by killing the bacteria responsible. His most famous invention, the Finsen lamp, separates UV light from other wavelengths and has been used to treat a variety of skin diseases.[27] A bulky cylinder with four telescopes awkwardly protruding from its base, the original Finsen lamp resembles a clunky Soviet-era satellite, but Finsen's work opened up the new world of phototherapy. In 1903, he became the first Scandinavian to be awarded the Nobel Prize for Medicine.

Today, the most commonly used wavelength to treat skin diseases is that of UVB light. UVA light can also be used, in a process called PUVA. In PUVA, the patient is exposed to UVA light after taking a tablet containing psoralen, a molecule derived from natural plant compounds that makes the skin more sensitive to light, so that the UVA can be used at lower doses. Phototherapy is particularly effective at treating psoriasis, in which the keratinocyte cells of the epidermis proliferate excessively (being replaced every five days compared to the usual thirty), causing the condition's characteristic scales and plaques. In phototherapy, UV light damages the DNA of these cells, which stops them proliferating. In fact, the damage caused by phototherapy probably affects most immune cells in the skin, as shown by the effectiveness of sunlight and phototherapy on diseases where various immune cells are overactive, including eczema and cutaneous T cell lymphoma. Phototherapy also causes melanocytes to produce more of the dark, protective pigment melanin, and it is sometimes effective in darkening the white patches of vitiligo.

While skin can be healed by powerful UV light in a controlled setting, it may come as a surprise that simple, visible light has left an equally impressive mark on healthcare. On a warm summer's afternoon in 1956, in the courtyard of a local hospital in Rochford, an unassuming town in Essex, Sister Jean Ward was about to contribute to one of the greatest discoveries in the history of paediatrics. Like Dr Finsen, she was a sun-lover and would take premature infants out into the hospital courtyard: 'A combination of fresh air and warm sunshine would do them more good,' she would say, 'than the stuffy, overheated atmosphere of an incubator!'

On a ward round, another nurse noticed that one of the premature babies, whose skin was initially yellow with jaundice, returned to a healthy pink colour within a couple of days, but still had an odd, well-demarcated triangle of jaundiced skin. It so happened that this child had been exposed to sunlight but the area of yellow skin had been covered by the corner of a sheet. Neonatal jaundice, where the skin of an infant is yellowed by the build-up of bilirubin – the pigment released during the breakdown of red blood cells – is often harmless and resolves within a few days, but it can interfere with sleep and feeding patterns and, in some untreated cases, lead to brain damage.

A couple of weeks after the yellow-triangle observation, and completely separately, the resident doctor on the ward, Richard

Cremer, noticed that a blood sample from one of the jaundiced infants, who was undergoing an exchange blood transfusion, had turned green after it had been left in direct sunlight on a window ledge. The levels of bilirubin in this sample turned out to be much lower than expected. It began to dawn on the team that something in sunlight could have a direct effect on bilirubin. Dr Cremer started to test the effect of visible light on bilirubin levels in blood, experimenting with different sources, including a local street lamp. He eventually discovered that blue light broke down bilirubin molecules and completely treated neonatal jaundice without the need for blood transfusions. High levels of blue light switch insoluble bilirubin into a soluble form that can be easily excreted from the body. It is now seen as one of the most significant findings in twentieth-century paediatrics, but at the time many in the medical profession would not believe that simple sunlight could have such a transformative effect on a specific condition.[28] It took another thirteen years before Dr Jerold Lucey's team at the University of Vermont confirmed Dr Cremer's findings and light therapy for neonatal jaundice became standard practice.[29]

The discovery of light therapy is a classic case of a stroke of serendipity completely transforming medicine, improving and saving countless lives. Even so, the first visit to a neonatal ward can feel slightly surreal, with all the tiny premature babies bathing in deep blue light as though under the tractor beam of a UFO. Currently, a team in Switzerland is developing illuminated pyjamas for jaundiced infants so they can be exposed to this shortwave light while wrapped up and in their mother's arms.[30]

In adults it is clear that a number of skin conditions are improved by UV light, but what about visible wavelengths of light? In 2016 a photo of reality-TV star Kourtney Kardashian, face obscured by a slightly terrifying white mask emitting a deep-blue glow, was seen by her 36 million Instagram followers and rapidly brought LED-light therapy into the public arena. The proponents of LED therapy, from beauticians to a long list of Hollywood

A-listers, claim that it can treat anything from acne pimples to age-related wrinkles. The theory is that wavelengths of 'high-energy' blue and purple light are able to kill off *Propionibacterium acnes* bacteria, one of the causes of acne, while 'softer' red and pink light is supposed to speed up healing and slow down ageing. However, it currently looks as though visible-light therapy is more con than cure. While it is certainly true that high concentrations of blue light do kill certain bacteria in the laboratory, there is no evidence that it can treat acne. A systematic review, which statistically combined the effectiveness of seventy-one studies testing light therapy on acne, found that there is currently no high-quality evidence that either blue- or red-light therapy works.[31] This could change in the future, but for the moment there are other more effective and cheaper treatments. It is an inconvenient truth that the success stories of novel alternative treatments we see in magazines and newspapers – often influenced by the companies that make millions from them – do not stack up against hard evidence. Nowhere is this truer than the care of our skin, where the insecurity we feel about our appearance feeds a multi-billion-dollar industry.

One procedure for acne that goes a step further than LED-light therapy shows more promise. In photodynamic therapy (PDT), a photosensitizer such as aminolevulinic acid is applied to the skin beforehand. The chemical is absorbed by the blocked and damaged pores of acne, and when light therapy is applied *Cutibacterium acnes* bacteria are destroyed. Research into blue light is also shedding light on the skin's relationship with the sun, and it continues to surprise. In 2018 a group of Canadian scientists from the University of Alberta discovered that visible blue light emitted from the sun could explain our winter weight gain.[32] While high-energy UVB light stops at the epidermis to do its damage and UVA penetrates deeper into the dermis, this study found that the sun's visible blue light wavelengths are able to dive through these layers into the fat cells, called adipocytes, of the hypodermis. When blue light strikes the adipocytes, it shrinks the size of these lipid droplets, reducing the amount of fat these cells can store. This could help explain why we tend to put on those insulating pounds over the short, dark winter days, even when taking Christmas calories into account. It is well established that the detection of light by human eyes influences circadian rhythms and their metabolic effects on the body, such as increasing the release of the hormone cortisol in the morning to boost blood glucose levels. But it could be that our skin also influences our seasonal rhythms.

It is clear, then, that the sun's rays can both harm and heal. In modern discussion, nowhere is this dual nature more apparent than with the confusion over vitamin D. Although sunlight damages our skin, it also supplies most of our vitamin D needs, and sun avoidance can lead to vitamin D deficiency. In Jordan, one of the sunniest countries in the world, four-fifths of all women – many of whom have skin hidden from the sun under Islamic dress – have vitamin D deficiency, compared to less than one-fifth of men.[33] Vitamin D is unique amongst essential nutrients in that most of our needs are met through skin, not diet. To add to the confusion, the active form of vitamin D is not actually a vitamin but a hormone. It plays a crucial role in the body's regulation of calcium, phosphate and other minerals. Diseases caused by vitamin D deficiency demonstrate its importance in supporting bone mineralization: in osteomalacia (and rickets, its childhood form) bones are softened and easily bend and break, and muscles are weakened. Its effects are not limited to bones and muscles; there are receptors for vitamin D in almost every cell of the body, and there is emerging evidence that this hormone may influence the immune system, cancer protection and even mental health. While it is very important to avoid vitamin D deficiency, this molecule may not be the sole panacea for all ills that some claim it to be; the overall evidence for the effects of vitamin D supplementation on heart disease, diabetes and cancer is very inconsistent.[34]

Skin is a vitamin D factory where UVB, the solar sword also responsible for DNA damage and cancer, creates the vitamin in a two-stage process. First, UV rays split apart a precursor molecule (called 7-dehydrocholesterol) in the skin into previtamin D3. This is immediately further broken down by heat into vitamin D3. This molecule then travels to the liver and kidneys where it is transformed into active vitamin D, which goes on to carry out its important functions around the body. Vitamin D can also be obtained through diet. Foods high in this vitamin include oily fish and fortified dairy products, but it is tricky to get enough vitamin D through diet alone. Achieving an adequate daily dose of vitamin D without any sunlight would almost certainly require supplementation through tablets.

If you have ever thought that our dual consumption of this vitamin is slightly bizarre, we are outdone by our pets, which also use the skin and diet to obtain the sunshine vitamin, but in a curiously different way. Both cats and dogs secrete cholesterol-containing oils from their skin on to their hairs. When exposed to sunlight, the cholesterol compounds in the oil are converted to vitamin D. But it can now only enter the animal's body orally, and so this is one of the reasons for your pet's constant hair-licking. It is likely that this seemingly roundabout way of obtaining vitamin D in some mammals is due to the thick layer of fur separating their skin from the sun.

Huge numbers of people across the world are vitamin D deficient while the global skin cancer burden is swelling at an unprecedented rate. In balancing our need for vitamin D with the damage the sun can do to our skin, the all-important question is how much sunshine should we let in? For most of the year it is certainly possible to obtain one's vitamin D requirements solely through the skin, one advantage being that there is no possibility of overdose, as the skin removes excess vitamin D.[35] For those living in Northern Europe and the northerly states of the USA, adequate intake can be achieved through exposing forearms, hands and legs to the sun for ten to thirty minutes per day (or approximately half of the time it usually takes for one's skin to redden) between 11 a.m. and 3 p.m., two to three times a week between April and September. There are two important caveats to go with this advice, however. Firstly, this regime is dependent on many variables, including: latitude, cloud cover, air pollution, skin pigmentation, clothing, sunscreen use and – importantly with humans – memory and discipline. Secondly, it is important to note that even these short periods of time can lead to DNA damage that cumulatively can lead to skin cancer. As with the bad news that there is no such thing as a healthy tan, there is no agreed 'safe' limit of sun exposure.

The American Academy of Dermatology, supported by many other medical bodies worldwide, recommends that we 'don't seek the sun'.[36] As it is also possible to obtain our vitamin D requirements through diet and supplementation, guaranteeing an adequate vitamin D intake without any risk of skin cancer, taking regular supplements is a rational approach. The American Institute of Medicine recommends a daily Vitamin D supplement of 400IU (International Units) for infants under the age of one, 600IU between the ages of one and seventy and 800IU for those over seventy. Meanwhile, the UK Scientific Advisory Commission recommends a year-round daily 400IU of Vitamin D from natural and fortified foods as well as supplements. The sensible answer to the vitamin D/sun damage conundrum probably lies somewhere in the middle: we can obtain enough of this vital vitamin through a combination of food, supplements and *protected* sun exposure. We should spend time outside every day to achieve personal happiness, rest and exercise, but there is no need to actively seek the sun to 'top up' vitamin D levels. Avoiding tanning and burning is extremely important and vitamin D dietary supplementation is safe and beneficial to health, particularly for the many who have a deficiency. Our skin, like Icarus's father, teaches us to take the middle way: to neither fly too close to, nor too far from, the sun.

5

Ageing Skin

Wrinkles and the war on mortality

'Time heals all wounds.'
ANON
'Time wounds all heels.'
DOROTHY PARKER

I MET NANCY JUST once, on a short visit to a hospice. She lay on a bed by the window in the corner of the room, propped up by two pillows. Her forearms, essentially bones thinly wrapped in blotchy white-and-purple skin, rested over her cotton blanket. The skin covering her face was fragile, sunken into her cheeks, creating canyons of wrinkles. It was one of my first weeks as a medical student, and I had jumped at the opportunity to escape the textbooks and see 'real' patients. Even with no medical training at this point, I could sense that something was very wrong. The general practitioner I was shadowing nudged me forward to examine Nancy.

'Hello, Mrs Wood. Can I listen to your heart, please?'

'Yes, of course, dear,' she mumbled, turning her glazed eyes towards me and breaking into a gentle smile.

As I leaned forward and strained to listen to Nancy's slow, faint heartbeats, I was thrown by another sense: up close, she smelled *really* bad. The nurse and doctor decided to have a closer inspection of her abdomen and the skin on her legs. It was only when she was gently turned over that we all saw the problem. At the base of Nancy's back, right above her coccyx (tailbone) was a small, perfectly circular ulcer. With red, inflamed edges and a cavity seeming to tunnel through various layers of tissue, it was as though someone had taken an oversized hole-punch to her back. A small amount of pus was seeping from the opening, which had left a damp, sticky patch on her bed sheet. The pressure from lying largely in one position in bed for days on end had obstructed the blood vessels that supplied the area of skin above her coccyx. Without a supply of oxygen and nutrients, her thin, delicate skin had begun to die. This skin around the edges of the ulcer would also shear apart when she sat up in bed and was transferred from bed to wheelchair and back, widening the wound.

When I spoke to the doctor again a few weeks later, I learned that Nancy had died.

'It wasn't the ulcer that hastened her death,' he said, after I assumed it had played a role. 'Your *Gestalt*, your gut feeling linking your perceptions to your subconscious, told you that she was dying, but the clue was on her skin long before we saw the ulcer. Did you see the marbled, mottled purple pattern along her arms? It's the harbinger of death. It shows that her blood circulation is collapsing and, although it's not an exact science, it often begins to appear about a week before death.'

Our skin ages with us, telling our story, for better or worse. Seven in ten elderly people in Britain have a skin disease, from the itching of scabies and venous eczema to life-threatening skin cancers. Nancy's humble pressure sore, often dismissed as prosaic by medical students, is also found on the skin of as many as 30 per cent of care-home patients. Fiendishly difficult to treat, these ulcers cause untold pain and misery, and infections can even lead to death. The cost of hospital stays, countless bandages and antibiotic regimens for pressure sores exceeds £4 billion annually in the UK alone.[1] Reports also find that elderly patients are often too ashamed to talk about their skin symptoms, and this unromantic corner of medicine has long been neglected.

When we hear the phrase 'anti-ageing', we don't tend to

conjure up an image of new treatments for arthritis, dementia or hearing loss; we think about our skin. Our appearance is a critical part of our being, even more so than the risk of death, and – as we saw in Chapter 4 – people are much more likely to wear sunscreen if they are told it slows down skin-ageing than they are for preventing life-threatening skin cancer. In Aldous Huxley's 1931 novel *Brave New World*, the citizens of World State are artificially kept perpetually young, with no one noticeably ageing past thirty.[2] Linda, a woman from the uncivilized West, arrives in World State to the horror of its residents:

> 'Bloated, sagging, and among those firm youthful bodies, those undistorted faces, a strange and terrifying thought of middle-agedness, Linda advanced into the room, coquettishly smiling her broken and discoloured smile.'

Today's obsession with anti-ageing and advancements in its practice, from creams to cosmetic surgery, reveals Huxley's dystopian prophecy to be at least partially fulfilled. Even though modern medicine has enabled us to push the definition of 'old age' further and further back, the autumn and winter of life – traditionally associated with wisdom and reverence, which in some cultures is still the case – are increasingly viewed with fear in Western society. But by filling in our wrinkles, do we also wipe out the positive aspects of old age?

Fuelled by a modern cult of youth and a $400 billion cosmetics industry, a visible battle against mortality plays out on the skin of millions. A glut of flawless complexions is thrown at us every day through television, while social media reinforces that, in our uncertain world, we can control our destiny by controlling our appearance. It would be easy to say that our anti-ageing culture is unhealthy and we should all simply embrace our wrinkles, but it's not as simple as that. Our skin is our self, and changing the way we see our skin is essentially changing part of our very being. Our society reveres the young, so the desire to keep the skin as youthful and healthy-looking as possible is understandable and it can be very difficult to see the positives of old age. For those to whom it matters, there are scientifically proven ways to slow down the signs of ageing on skin, and you can take away all, some or none of the anti-ageing science in this chapter.

Unless you are in your teens, you will at least be aware of your skin's inevitable wrinkly path, and many of us will be taking active steps to fight against this process. But what do we actually know about how the skin ages?

The completely unstoppable form of ageing is what comes with the march of time, and is known as 'intrinsic' or 'chronological' ageing. As we age, a number of changes occur in our skin. Our outer epidermis takes longer than the usual thirty to forty days to replace skin cells, and the layer connecting the epidermis and dermis begins to flatten, thinning the skin. The most important changes, however, take place in the deeper dermis. Our fibroblast 'builder' cells begin to think about retirement and start to slow down their production of collagen (the protein that gives skin its strength and plumpness), elastin (the protein that brings our skin back into shape after stretching) and glycosaminoglycans (the molecules that attract water and lubricate the skin). One particularly sobering statistic is that from our early twenties we start to lose roughly 1 per cent of our skin collagen a year, and this accelerates after the age of forty. Like an overworked oil field on an arid landscape, sweat and oil glands also begin to dry up. Late on in life, the blood vessel walls in our skin begin to thin, leading to easy bruising. The gradual loss of fat below the skin compounds the collapse of skin shape and sunken facial features. Overall, skin begins to lose its thickness, plumpness, elasticity and hydration.

Rates of intrinsic ageing vary between sexes, races and even families. Our skin has various oestrogen receptors that help induce the production of collagen and hyaluronic acid which hold water in the skin. So the drop in the level of sex hormones in the menopause pushes a foot down hard on the ageing accelerator in women. When it comes to skin colour, there is truth behind the old saying 'black don't crack'.[3] Black skin tends to have a higher level of lipids and protective melanin, meaning that, on average, black skin ages best, with white Caucasian skin coming last in the rankings. The complex concoction that is our individual genetic make-up also influences the way we age, probably in many ways we do not yet know. Intrinsic ageing even varies on the skin of one individual. Thinner areas of skin, such as the eyelids, start to age first. At some point our final enemy in intrinsic ageing, gravity, makes sure it wins the battle over drooping and sagging. Research is continually uncovering new mechanisms of intrinsic ageing, with the hope of potentially slowing it down. A 2018 study at the University of California, San Diego, found that some fibroblasts in the dermis have the ability to change into fat cells, which help give skin its youthful plumpness.[4] As we age, the ability of fibroblasts to transform into fat cells diminishes. Intriguingly, the protein that blocks this transformation process, transforming growth factor beta (TGFβ), also stops fibroblasts producing antimicrobial molecules. This may help explain why the elderly are more susceptible to skin infections, and a potential treatment that blocks TGFβ could be both beautifying and bactericidal.

Before diving into how to tackle the cracks and crevices that form on our face as we age, it's useful to look at wrinkle nomenclature. Deep lines (or furrows) usually start as 'dynamic' and end up 'static': when teenagers smile, 'dynamic' lines briefly appear on the outsides of their eyes before rapidly disappearing. Over time, these become static 'crow's feet'. Meanwhile, fine lines are usually related to irregular thickening of the skin combined with water loss, and these are associated with what we call 'extrinsic' factors. All our organs undergo largely unpreventable chronological ageing, but our skin is hit twice; delicately perched on the exterior of our body, it is exposed to the environment. If we are to push back against the furrowing of our outer layer, we need to be aware of the environmental accelerators of ageing.

To explore the multifaceted physical assault on our skin during the day, let's chart a day in the life of a wrinkle. You get out of bed, go to the bathroom, get dressed, then have breakfast. As you leave the house you come face to face with the greatest cause of skin ageing: the sun.

As a fresh-faced medical student, I vividly remember seeing a mother and daughter in the clinic. A couple of minutes into the consultation, I asked the woman with the wrinkled, leathery and blotchy skin, 'So, is Stephanie your only daughter?' After a couple of seconds of confused silence (which soon became awkward silence), I realized that the daughter, in her early forties, looked considerably older than her sixty-year-old mother. It transpired that the daughter had used sunbeds for the best part of thirty years and spent time on the Spanish coast whenever she could. Her mother hadn't actively avoided the sun during her life, but had never sought it out. The patients I see who appear older than they are have usually spent a lot of time exposed to the sun: gardeners, labourers and soldiers. Either that or they have regularly used sunbeds or 'made the most' of holidays on the beach. Sun-worshippers who used tan-increasing oils in the seventies and eighties often have the deepest cutaneous creases today.

Ultraviolet B light is the main contributor to sunburn and, eventually, skin cancer, but when it comes to skin-ageing, we need to watch out for its underestimated partner, ultraviolet A. These rays are weaker but can travel further than UVB rays, digging deeper into the all-important support structures in the extracellular matrix of the dermis. UVA damages the dermis by causing inflammatory pathways that lead to the release of molecules called matrix metalloproteinases.[5] These go on to break down our precious supply of skin collagen and – as if it wasn't enough to lose 1 per cent of our collagen each year – slow down the rate of collagen synthesis by fibroblasts. UVA also dilates and breaks down blood vessels in the dermis, resulting in the small 'spider veins' usually visible on the nose and cheeks. Other important damage includes the destruction of retinoic acid receptors, contributing to a deficiency of vitamin A in the skin. The critical importance of UVA in this photoageing process is that we do not need sunburn, or even noticeable tanning, for it to inflict age damage on our skin.

UVA is also able to penetrate glass, while UVB cannot, so while you're unlikely to get sunburn through a window, the skin-ageing effect continues as long as you are exposed to sunlight. It is not uncommon for elderly truckers traversing the American Midwest to have one side of their face drooping and wrinkled, while the other half looks two decades younger. Unlike intrinsic ageing, sun damage unevenly thickens our skin, bringing about the excessive proliferation and mutation of skin cells, shown in a variety of common precancerous skin lesions (called actinic or solar keratoses) and skin cancers, many of which are directly sun-related. The leathery, wrinkly, thickened skin caused by sun damage is the result of fibrosis, where photoageing is essentially an extremely slow healing response. Wrinkles are literally the scars of accelerated ageing.

Also heralding ageing skin are old-age, or liver, spots. These dark-brown blotches are often said to be the cause of our hands giving away our age. The name 'age spot' is somewhat misleading, however, as the marks are directly related to sun exposure, not to age. In areas of skin that have been regularly exposed to UV light, namely our face and hands, overworked melanocytes have been producing so much excess melanin that patches of pigmentation on the skin become permanent.

Sunlight is certainly the greatest contributor to skin-ageing, probably more than all other factors combined, including time itself. The key to youthful skin is sun protection, and the most effective anti-ageing cream is sunscreen.

After your encounter with the sun, you arrive at work and power up your computer. For most of the rest of the day your face is 30cm away from a glowing source of artificial light. Some are now arguing that HEV (high-energy visible) blue light, which is emitted by normal sunlight and by the LED displays of computers and smartphones, can contribute to accelerated wrinkling. Could these devices that we now completely rely on be making our faces look older? Current sun creams, which block out only UV light, would have no effect on screen light sagging our skin. Dermatologists are currently debating whether to include HEV protection, but the jury is still out.[6] There is a trace of evidence to suggest that HEV increases the collagen-eating matrix metalloproteinases, but there's absolutely no reason to suggest that computers cause crow's feet, to any significant extent, on human skin, and they certainly don't cause cancer.

It's now the lunch break and you go down to the canteen. There's almost no natural light, so surely you can escape wrinkles here? Don't be so certain. Sugars in our diet bind to proteins to form AGEs (advanced glycation end products), which attach to collagen, making it brittle. There is evidence that AGE deposition, which is particularly high in certain conditions related to blood sugar, such as diabetes, contributes to skin-stiffening, loss of elasticity and an increase in pigmentation.[7] We don't know to what extent sugar ages our skin, but there are lots of other reasons to limit sugar in our diet. The Western obsession with seeking low-fat foods, while simultaneously turning a blind eye to refined carbohydrates, is actually the opposite of what our skin needs. We need a balanced diet, complemented by adequate protein intake to support our skin and hair. Fruit and vegetables, particularly colourful ones, have repeatedly been shown to be good for the health of our largest organ, whether it is by directly combatting oxidative stress (the accumulation of tissue-damaging molecules called free radicals) or by influencing the slower, serpentine routes indirectly influencing our skin's health – by boosting our immune system, for example. You can use the most expensive anti-ageing creams and see a dermatologist once a week, but a poor diet will always show through your skin.

By now it is early afternoon, and you head back to your desk. Mental stress can also affect the skin's appearance, as we'll see in

Chapter 7, and perhaps you're considering a cigarette break before too long? Alongside sun damage, smoking is an incredibly powerful accelerator of skin-ageing. After just a few years of smoking, premature wrinkles will start to appear and the skin takes on a dull, sallow complexion.[8] This is abundantly clear in photos of twins who live relatively similar lives except one smokes and the other doesn't. Some of the 4,000 chemicals in cigarette smoke increase matrix metalloproteinases, which damage collagen and elastin, while nicotine causes the narrowing of blood vessels in the skin, reducing its oxygen and nutrient supply. If the other health benefits of stopping smoking are not apparent, or don't matter to you, it's indisputable that you will look more healthy and youthful after quitting, and it's never too late to quit.

What about the repeated pursing of lips when smoking? As a child, whenever I screwed my face up during a sulk or tantrum (and this was rather more often than my parents would have liked) my grandmother would say, 'Don't make that face – if the wind changes you'll stay like that for ever!' Some rudimentary experimentation showed me that this wasn't the case, but the movements we make with our face over decades do in fact become set in stone. So how far should one go in reducing or stopping these movements for cosmetic benefit? There are plenty of articles in fashion magazines that suggest ways in which to reduce your facial expressions when you are happy, frustrated or quizzical, but is it worth suppressing our emotions, and the intricate way our skin communicates them to others, just to slow down the onset of frown lines? This hints at the fundamental question when it comes to our ageing skin: what is the point in preserving the youthfulness of your visible self if you do not actually live while you're living? In itself, trying to preserve youth and beauty in skin is understandable, but if you are actively trying not to express human emotions to achieve it, you have probably crossed a wrinkly line. If only we could all look at it as poetically as Jimmy Buffet, who sang 'wrinkles only go where the smiles have been'.[9]

Now, at the end of the day, work has finished and you burst through the office doors on to the street. It's the middle of rush hour and the haze of car fumes lingers in the air like a grainy, translucent mist. Like our lungs, human skin has not adapted to cope in an environment of fumes and toxins. There's not a huge amount of evidence to show that 'city skin' is a thing, but some science is starting to suggest that some compounds in air pollution, such as nitrogen dioxide (NO_2), do in fact contribute to wrinkling.[10] We now know that these toxins can enter our skin, produce free radicals and set off inflammatory cascades. And these pollutants are everywhere. For context, London's Oxford Street broke its yearly limit of NO_2 within the first week of 2017.[11]

You arrive home. You have dinner, wash and are ready for bed. Surely there's nothing else that can go wrong in this wrinkly story? Wrong. Physical pressure can make marks on the skin, and the effects are more noticeable with age. Dermatologist Debra Jaliman argues that repeatedly crunching the side of your face against a pillow causes 'sleep lines'[12] and some beauticians and dermatologists I have spoken to claim they are able to tell which side of the face their client lays on the pillow. These marks are temporary, but if they are visible for most of the day – when others are seeing them – for many they may as well be permanent. If sleep lines are a genuine concern, the best fixes are to sleep on your back with the help of a U-shaped pillow, or to use pillows designed to help reduce these lines; many still swear by silk.

There is also science behind the concept of 'beauty sleep'. A Swedish study in 2010 demonstrated that sleep-deprived people appear less healthy and less attractive than the well-rested.[13] The same team later found that the others perceived sleep-deprived people as having noticeably changed skin: 'darker circles under the eyes, paler skin, more wrinkles/fine lines'.[14] A 2015 analysis of the skin of the chronically sleep-deprived also revealed a weakened skin barrier and increased signs of intrinsic ageing.[15] Inadequate sleep harms our immunological, metabolic and mental health, which inevitably accelerates skin-ageing.

When it comes to reducing the onset of cutaneous contours, a sensible way to start is by addressing the environmental, 'extrinsic' factors of sunlight, smoking, diet and sleep, of which the most important is sunlight. But is there anything that we can add in to our daily routine to slow down – and maybe even reverse – the wrinkly tide of age?

When it comes to anti-ageing creams, the mass of competing advertisements in our shops and supermarkets and on television can be pretty overwhelming, each making claims that their product 're-plumps' wrinkles and 'lifts' firm skin, while the perennially ambiguous 'rejuvenates' is another favourite promise. It also seems that, on roughly a three-month rota, a new celebrity finds the secret to age reversal, from cryogenic chambers to Kim Kardashian's 'vampire facials', where one's own blood is drawn and centrifuged to divide the red blood cells from the plasma, the latter being reapplied to your skin, which has been pre-pocked by microneedles. Fad facials are nothing new, from the ancient Romans bathing in crocodile dung to the fifteenth-century serial killer Elizabeth Bathory allegedly covering herself in the blood of her virgin victims to retain her youth. My personal favourite is the choice of the nineteenth century's Empress Elisabeth of Austria, who preferred 'Crème céleste', a concoction of spermaceti wax (found in the head of sperm whales), almond oil and rosewater, and who would sleep with raw veal on her face, with crushed strawberries thrown in for good measure, held together by a made-to-measure leather mask.

We laugh or recoil at these eccentric cosmetics, yet we are just as susceptible and have to some extent been brainwashed by the beauty industry to think that the more expensive or dramatic the treatment, the better or more efficacious it is. This simply isn't the case, with some studies suggesting that cheap moisturizers have exactly the same effect as their expensive, 'anti-ageing' counterparts.[16] Luxury anti-ageing creams also exploit one of the many blind spots in human psychology. You walk into a department store and are looking along a shelf dedicated to anti-wrinkle creams. Two pots produced by different companies sit next to each other; one looks slightly bland but is reasonably priced while the other looks smart, glossy and fresh out of the laboratory but costs five times more. We're tempted to go out of our price range to buy the expensive pot because it taps into our insecurities. Luxury cosmetics lower our self-esteem by creating the illusion of a higher level of beauty, only attainable through exclusive products, and we feel a need to make up for the perceived gap between the skin we think we have and the skin we want. The product actually drives us to buy it by making us feel worse about ourselves. Charles Revson, one of the founders of the modern cosmetics industry, was telling the truth when he said, 'In the factory we make cosmetics. In the store we sell hope.'

So you've picked up the glossy bottle, but what about the claims on the packaging? In the food industry, if you want to make a health claim about an ingredient, it must (quite rightly) be backed by a body of scientific evidence. Such regulation, in the UK at least, does not apply to skin products. Evidently misleading claims can be challenged by the Advertising Standards Agency, but crafty cosmetics companies can easily be economical with the truth. The packaging of the cream declares that it's 'clinically proven to reduce wrinkles'. This could be true, but a 'clinical' change could be one seen only under the microscope, and never visible to the human eye. But it's been 'dermatologically tested'? In theory it could just have been tried out for a few days on the skin of just one participant, perhaps the marketing director's indifferent father. If the product has a list of 'active ingredients', these could simply have been tested *in vitro* (in the laboratory) and their effects may never have been observed on human skin.

We are still some way from an elixir. While some products genuinely have ingredients that can slow down the visible effects of ageing (such as sunscreens and retinoic acid) and some are able to make their customers feel content and confident – and that's certainly worth something – when you are spending money or seeking the truth, a bit of healthy scepticism never goes amiss.

But even the eye-watering prices of today's anti-ageing skin creams pale into insignificance compared to the rejuvenation treatments used by Queen Cleopatra of Egypt. She had a stable of seven hundred donkeys, which supplied her daily bath of milk. Although this may sound as patently ridiculous as the veal facials of Empress Elisabeth, Cleopatra was perhaps on to something. Milk contains alpha hydroxy acids (AHAs), of which glycolic acid is still particularly popular today as an ingredient of skin creams. These acids promote skin exfoliation and the renewal of epidermal skin cells, but whether they are also able to penetrate and firm up the dermis is harder to know. When it comes to exfoliation – removing the outer layer of dead skin cells – dermatologists generally agree that doing this once or twice a week, scrubbing the skin with gentle pressure, is enough. It takes considerable effort for our skin to create an effective barrier to the irritants and infective agents of the outside world, and we don't want to rub it off.

The science from Cleopatra's stables suggests that there may be some molecules that really do have a positive effect on ageing. The ingredient with the most scientific evidence behind it – indeed, a number of dermatologists believe it is the only one with any evidence at all – is retinoic acid. This is a breakdown product of vitamin A, which is vital for skin and body health and is sourced from beta-carotenes, found in colourful vegetables such as carrots. It is part of the family of retinoids, compounds chemically related to vitamin A. In 1960 Albert Kligman found that a derivative called tretinoin (which he branded Retin A) was incredibly effective at treating moderate and severe acne.[17] About a decade later he realized that retinoic acid had another, even more lucrative, potential: it increases collagen synthesis, thickens the dermis and exfoliates the outer epidermis, noticeably smoothing out wrinkles. The way Kligman went about making his discoveries, however, was certainly less than ideal and has helped form our current medico-legal laws surrounding consent. From the 1950s to the 1970s he carried out a series of cosmetic dermatology experiments on the inmates of Holmesburg Prison in Philadelphia. When he entered the prison for the first time, Kligman commented that, 'All I saw before me were acres of skin . . . it was like a farmer seeing a fertile field for the first time.'[18] His first venture into the prison was to treat an outbreak of athlete's foot amongst the inmates. A corruption of his position of trust and the vulnerability of the inmates led to him slowly starting to expose them to skin infections, and even going beyond his speciality to test psychoactive drugs on the prisoners.

Retinoic acid thins the outer layer of our skin by roughly a third, which means it slightly drops our skin's natural SPF, making it more vulnerable to sunburn. It is therefore generally recommended that retinoids be applied at bedtime, before the skin begins to repair itself overnight. One particular retinoid has also caused considerable controversy, however. Retinyl palmitate is included in many sunscreens, allowing manufacturers to claim that their products also have anti-ageing properties. Unfortunately, retinyl palmitate is not only one of the least effective retinoids for flattening wrinkles, but has also been associated with skin cancer. Some studies run by the National Toxicology Program in the USA found that mice swabbed with retinyl palmitate had higher rates of skin cancer than control groups.[19] In 2010, the non-profit Environmental Working Group recommended consumers avoid this product in sunscreens.[20] The results of the studies have been heavily contended by scientists and dermatologists, a debate that has not been helped by the fact that some of those involved have financial interests in cosmetics companies. One thing we can be certain about is that sunscreens definitely do not cause cancer, but during the day it may be worth avoiding products that contain retinyl palmitate. Dermatologist and retinoid specialist Leslie

Baumann argues that, whether or not it causes cancer, there are in any case better retinoids out there: 'I do not feel that there is enough evidence to prove that it causes skin cancer. But, then again, can you give me one good reason to use it?'[21]

After sunscreens, retinoic acid is arguably the only anti-ageing cream that has robust evidence behind its claims. But it's not to be used too enthusiastically. A pea-size blob should be enough to cover the facial skin, and, in a similar way to exercise, we should apply these creams at a walk before we can run: increase to the recommended dose over a few days. More retinoic acid than this doesn't mean fewer wrinkles: it just means more burning, stinging and redness on the skin.

There are thousands of preparations and formulas for anti-ageing creams, all of which fall into just a few categories. Many people sing the praises of antioxidants. There is some, albeit limited, scientific evidence for these marginally reducing the appearance of skin-ageing. Those with the most evidence include nicotinamide, vitamin C, vitamin E, selenium and coenzyme Q10. The main issues are that many of the vitamins used are unstable, have short-term effects and it's not certain how deeply they penetrate the skin. It is possible that newer preparations are addressing this problem, but there is as yet no robust scientific evidence to support the use of vitamins in anti-ageing. Another interesting and rapidly evolving area is synthetic proteins. These include palmitoyl pentapeptides, which are essentially proteins with fatty acids attached. Some studies show that they can penetrate through the epidermis and stimulate collagen production.[22] It's hard to isolate any miracle molecule, but creams that combine antioxidants, peptides and a combination of other chemicals have been shown to have a moderately positive effect on reducing wrinkles. Professor Chris Griffiths at the University of Manchester led a clinical trial in which 70 per cent of participants who applied a combination serum (No7 Protect & Perfect Beauty Serum) enjoyed significantly reduced wrinkles after one year compared to placebo.[23] But no matter what any cosmetics giant or magazine proclaims, there's no silver bullet – yet – that can rewind the years. Throw in environmental factors and our uncontrollable genetic code, and the truth about ageing skin becomes a much more complex issue than we would perhaps like to hear.

In 1895 Dr Émile van Ermengem, a student of the legendary bacteriologist Robert Koch, was called to investigate a horrifying scene at a funeral in Belgium.[24] As if a funeral wake isn't distressing enough already, halfway through the meal some thirty guests began to lose all of their facial expressions. Their eyelids then began to droop, some lost vision and others could no longer swallow their food, and began choking and retching on the floor. Three stopped breathing and eventually died, their chest muscles failing completely. After meticulously investigating the disaster, Dr van Ermengem discovered that the culprit was a bacterium that had been lurking in some dodgy ham and was later named *Clostridium botulinum*. This organism produces a neurotoxin that paralyses the body and at certain doses can be lethal.

A century after the fatal funeral, the married Canadian doctors Jean and Alistair Carruthers, an ophthalmologist and dermatologist respectively, found that patients whose eyelid twitching (blepharospasm) was treated with small doses of the botulism toxin were elated by its side effect: they didn't seem to age.[25] Their frown lines became frozen in time. Two decades after this semi-serendipitous discovery, the injection of botulism toxin (Botox) has become the world's most common cosmetic procedure. Botox paralyses facial muscles so that the furrows and frown lines of movement aren't visible. It's essentially the same strategy of the women of sixteenth-century Europe who caked their faces in a white paste made up of lead and vinegar (the classic image is of Elizabeth I); any facial movements would crack and ruin the make-up, so no facial movements were made. Since its increased popularity, Botox has become the subject of mockery, with so-called 'robotox' actors and newsreaders easy to identify due to their inability to produce even the smallest facial expression. As each year goes by, however, treatments more invasive than creams are becoming safer and more effective at appearing to slow the signs of ageing. Following in the tradition of Cleopatra's asinine baths, chemical and mechanical forms of exfoliation (using peels and microdermabrasion, respectively) are effective for some. Dermal fillers, usually composed of collagen or hyaluronic acids, the essential molecules of the dermis, do just what their name suggests: they fill in lines and wrinkles, and flatten out ageing skin, albeit only for a limited period of time. Other treatments use lasers and electromagnetic waves to sculpt and rejuvenate the skin. 'Radio frequency skin tightening' attempts to stimulate the growth of collagen and elastin in the dermis by heating dermal and subdermal tissue and letting the healing process remodel the supporting structures beneath our epidermis. This technique is also being used to break down the protrusions of fat in our skin, which we call cellulite, opening up another burgeoning area of aesthetic medicine.

Many creams seem to work wonders on some people but have absolutely no effect on others. It may be that celebrities, past and present, have stayed wrinkle-free through a combination of reduced sun exposure and personal genetics. Perhaps 'Maybe she's born with it' does contain a ring of truth. The bottom line is that if a cream (particularly one sold by an organization trying to make a profit) sounds too good to be true, it probably is.

Humanity is at war with ageing and skin is the ultimate battleground. With modern technology, however, it could one day be a war that we 'win'. In 2016 scientists at Harvard and Massachusetts Institute of Technology engineered a synthetic, wearable 'second skin' with a new technology that may visibly and naturally remove the appearance of furrows and blemishes.[26] In the war against wrinkles, the money spent on anti-ageing treatments would make even the US military blush. But if we win and are able to live like the citizens of Aldous Huxley's World State, perpetually looking thirty years old even as our inner organs begin to rot, is the cost really only financial? The people of Huxley's dystopian vision were taught to resent age and pretend that they were not going to die. Should wrinkles be 'cured' or should we be having a discussion in society about how we view age? In a world that likes to pretend decline and death don't exist, our skin urges us to confront our mortality.

6

The First Sense

The mechanics and magic of touch

'See how she leans her cheek upon her hand.
O, that I were a glove upon that hand
That I might touch that cheek!'

WILLIAM SHAKESPEARE, *ROMEO AND JULIET*

IF YOU ARE ever lucky enough to visit the Sistine Chapel in the Vatican, you will inevitably lift your eyes up to the heavens. The centrepiece of the magnificent ceiling is Michelangelo's *The Creation of Adam*, which must be one of the most compelling works of visual art in the world. It depicts God, enveloped in a shroud of angelic beings, extending his forefinger towards the limp, receptive hand of Adam, who reclines on the edge of the Earth. At first glance their fingers appear to make contact. A closer look reveals what it is that makes the painting so renowned: a tiny gap builds the electrifying tension and anticipation between Adam's hand and God's life-giving touch.

As this story of the skin moves from the physical towards the psychological and social, it is necessary to cross a tactile bridge. This bridge links the outside world – via receptors, nerves and brain tissue – to our mind, even our very being. Touch is the first of our senses to develop; it is certainly the most underrated and possibly the most remarkable. Glabrous (hairless) skin, which is mainly found on the fingers, palms and the soles of the feet, is populated by four groups of 'mechanoreceptors'. They all essentially respond to changes in pressure and distortion of the skin. Like a machine, they detect movement in the outside world and pass electrical information through individual nerves to the brain, and the body responds accordingly. Each of the four receptors is shaped by its function, each with strengths and weaknesses.[1] When they work together, something beautiful and near-miraculous happens. To fully appreciate and wonder at the complexity of our sense of touch, let's start by dissecting an everyday miracle: entering your own home.

You reach your hand into a pocket in search of the key you placed there when you left the house that morning. Unseeing, you rapidly make your way through mint wrappers, pens and loose change until you find the distinctive shape of the house key. How can you tell that you're feeling the key without seeing it? The receptors in our skin that detect the ridges, indents and shape of the key are called Merkel cells. These tiny, humble, circular discs are named after the German anatomist Friedrich Merkel, who called them *Tastzellen* (touch cells). Merkel cells are found in the basal layer of the epidermis, closer to the surface of the skin than the other three types of mechanoreceptors, and are highly abundant in our fingertips. They are able to detect infinitesimally tiny vibrations at low frequencies and can be activated by a mind-bogglingly minuscule application of pressure, detecting the displacement of 1 micrometre (0.001mm).[2] When the cell is stretched, sodium enters the cells from the extracellular fluid, initiating an electrical spike called an 'action potential', allowing a signal to be conducted along the nerve. Not only does this occur at the displacement of one thousandth of a millimetre but it also all takes place within one thousandth of a second. A Merkel cell is described as 'slowly adapting', which means it continues to send impulses to the brain for as long as there is a displacement in pressure on the skin's surface. This enables Merkel cells to continuously provide the brain with detailed information regarding the shape and edges of an object.

Now that you have found the edges of the key, you need to

grip it from both sides with exactly the right amount of force. You don't want it to slip through your fingers, but nor do you wish to squeeze it too hard (this delicacy of touch is probably more important when holding, say, your mother-in-law's expensive china plate). This balance is enabled by receptors called Meissner corpuscles (or tactile corpuscles), which are found slightly deeper than Merkel cells and have a rounder, bulbous form consisting of coiled, encapsulated nerve endings. These nerve endings sense vibrations, and unlike Merkel cells they are 'rapidly adapting', only recording the onset and offset of skin indentation. This is the reason why we feel our clothes when we put them on, but don't notice their presence for the rest of the day. The most impressive feature of Meissner corpuscles is that they literally catch us every time we fall. As you hold the key, it actually slips a thousandth of a millimetre a number of times a second. Our Meissner corpuscles can detect this loss and, in a series of rapid reflexes, cause our skin to tighten so that we don't end up dropping the object. All of this is completely subconscious.

You've managed to locate your keys with your Merkel cells and can hold them securely thanks to the rapidly adapting Meissner corpuscles. But now comes the Herculean challenge of successfully inserting the key into the lock. Enter Filippo Pacini. In 1831 this nineteen-year-old Italian medical student was dissecting a human hand when, paying extraordinary attention to detail, he unearthed some 1mm-long lumps in the skin. I have seen models of these beautiful skin receptors, 'Pacinian corpuscles', at the University of Florence's anatomy museum. These multi-layered structures reside deep in the dermis of the skin and somewhat resemble onions. When we feel the faintest pressure on our skin, the layers are squeezed together, deforming the corpuscle and firing off messages to the brain. Pacinian corpuscles have a long range for detecting pressure and vibrations, so much so that one of these tiny onion-like structures can locate a vibration anywhere in a finger. In fact, they can sense vibrations coming from anything our fingers are holding. These are absolutely critical to the humanness of the sense of touch; when we are holding tools we can 'feel' the actions at the working end of the tool as though it were an extension of our skin, whether it be a surgeon's scalpel or the key you are currently navigating into the lock.

Now that the key has made its way to the back of the lock, you need to turn it between thumb and forefinger before you can finally enter your home. This is made possible by your fourth and final skin mechanoreceptor, the Ruffini ending. Shaped like spindles and running parallel to the skin's surface, Ruffini endings are less concerned with skin indentation and instead detect horizontal stretching. Although these are less numerous than the other three mechanoreceptors, and we know less about the way our brain makes sense of Ruffini signals, it is likely that they both recognize stretch in the skin and respond to changes in the angle of the hand and joint position, enabling you to know where your hand is moving to as you twist the key in the lock.[3]

These little-known mechanoreceptors in the skin, named after two Germans and two Italians from the nineteenth century, bring about the minor miracle of manipulating objects. The ability to handle tools with such dexterity, and as though they are an extension of our own skin, is what sets us apart from animals and (at the time of writing) robots.

The remarkable reflexes of these 'fantastic four' mechanoreceptors enable touch to happen, but they do not explain how our brain knows where we are being touched. The physical reality of the outside world and the picture our brain creates to perceive this reality are in fact two very different things. Early explorers and cartographers tried to fathom the world and represent it visually in a way that could be understood by those at home. The mind understands the tactile world through two maps: the skin itself and the 'sensory homunculus' in the brain.

During the 1950s Dr Wilder Penfield, an outstanding Canadian neurosurgeon, was busy trying to treat patients with intractable

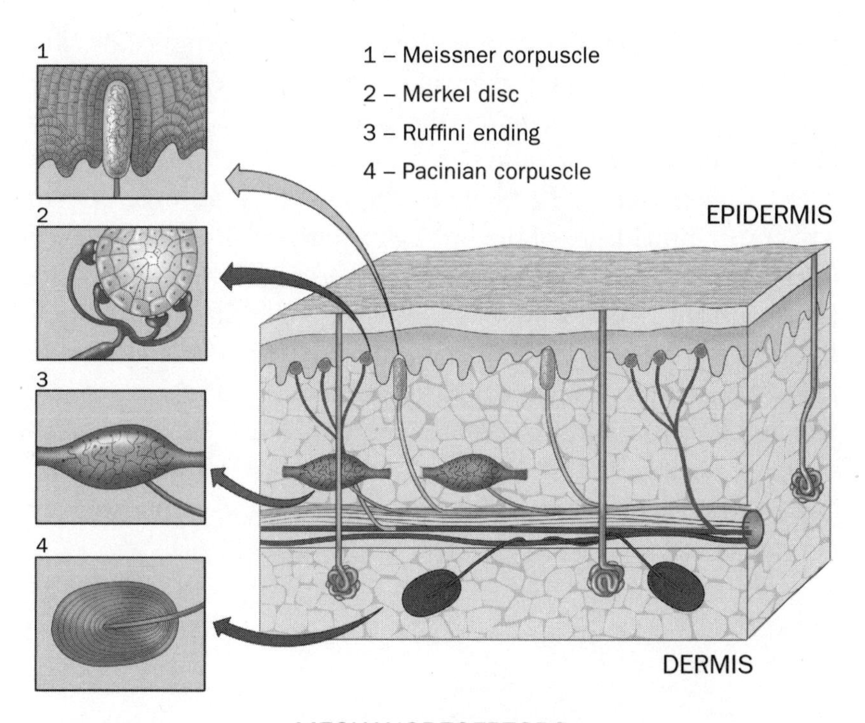

MECHANORECEPTORS

epilepsy.[4] Many of his patients suffering with epilepsy tended to experience auras: a sense that a seizure was about to start. He theorized that if he could remove a piece of skull and arouse the aura by touching an area of the patient's brain with an electrode while the patient was fully conscious, he could find the area of brain responsible for the seizure. While this experiment was only moderately successful, he stumbled upon something even more remarkable. When he prodded different parts of the surface of the brain during surgery, patients would feel sensations in different parts of their skin. Penfield then painstakingly recorded which areas of the brain corresponded with sensations on different areas of skin. Interestingly, the brain's map of the sensory skin seems jumbled up, and the amount of brain used is not related to the surface area of the body the skin covers. For example, the skin on the fingertip of our forefinger, with its considerable density of sensory receptors, requires the attention of a proportionally much larger area of our brain's 'skin map' than, say, the skin on our back. To represent this, Penfield created a model of a human where the appendages of the body were either shrunk or enlarged to represent the amount of space they occupy in the brain. The result was the sensory homunculus – a gangly, off-balance, 'grotesque creature' (as Penfield described it) with enlarged body parts where there is a high density of mechanoreceptors (hands, feet and lips), in contrast to those where there are far fewer (such as the torso and arms), which are spindly and diminutive. The brain's map of the body ever since has been called the sensory homunculus.

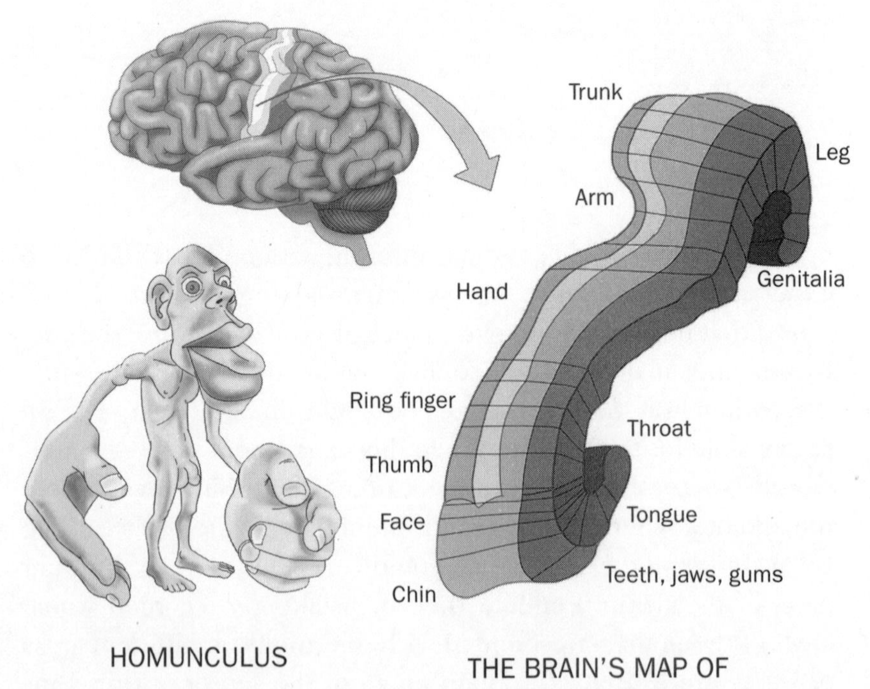

HOMUNCULUS

THE BRAIN'S MAP OF THE SKIN

The concentration of nerve endings in a given area of skin affects the accuracy of sensation. This explains the long-known fact that women tend to have a better sense of touch than men: just as the same amount of gin in less tonic makes a more potent aperitif, the same number of mechanoreceptors in smaller fingers makes a more sensitive instrument. As touch accuracy is inversely related to the size of fingers and hands, the more diminutive among us should take heart. Touch not only divides sexes and sizes, but also the ages. We begin to lose the receptors in our skin as we age, so the concentration slowly dwindles. This partially explains why the elderly are generally less adept at getting to grips with fine manual tasks. And while the loss of other senses – namely vision and balance – is commonly thought to be the cause of falls, a dwindling number of receptors in the skin of hands and feet is also to blame. But touch is not just down to how many receptors one has; it's also about how you use them. Blind people have been shown to have better discriminative sensory touch than people with sight and can translate the dots of Braille with extreme sensitivity and speed.[5] This is a demonstration of brain 'plasticity', where the brain essentially rewires itself to compensate for a lost sense; the skin can pick up the slack. One extraordinary example of the ability of our brain to change its wiring is the case of a thirty-six-year-old professor who suffered a stroke in the right side of her brain.[6] She lost almost all tactile sensation across the left half of her body and experienced 'hemispacial neglect', where she no longer had any awareness of her left field of vision. She would regularly bump into objects on her left-hand side, from doorways to passers-by. Thankfully these symptoms improved dramatically after eighteen months, but then she began to hear with her skin. Certain sounds, particularly the voice of a specific radio presenter, would unfailingly give her strong tingling sensations across the skin of her left hand. Brain scans revealed that there had been an anatomical reorganization of the neural connections in her brain. In the healing process following the stroke, neural connections had been formed between the auditory (hearing) and somatosensory (feeling) areas of her brain. This bizarre phenomenon is known as sound-touch synaesthesia – literally 'combined sensations'.

But the computer-like interface created by mechanoreceptors and neural connections is not the only way our skin enables us to touch. Have you ever wondered why your fingertips prune up in the bath? To a child it probably seems like the most impenetrable scientific mystery. It appears as though some kind of osmotic effect pulls small amounts of water out of the skin. But consider that this wrinkling only occurs on the glabrous skin of the palms, fingertips and feet. Surgeons in the 1930s also noticed that when the nerves to the fingers were cut, the wrinkle response vanished. It now appears that the wrinkling of our skin in water prepares us for an unusual touch challenge: gripping wet objects. In 2011 the neurobiologist Mark Changizi found that the pattern of wrinkling on fingers resembles a drainage network and that they indeed act like the treads of tyres.[7] In this discovery, the dynamic landscape of our skin forms a temporary mountain range, with watersheds to disperse new rivers. A year later a team at Newcastle University asked participants to dip their hands in warm water for thirty minutes and then complete a task of transferring wet marbles.[8] The subjects with fingers wrinkled from placing them in water were much faster at picking up and transferring these objects than participants with dry skin, but having wrinkled skin didn't improve the ability to pick up dry marbles. More recent studies, however, have shown results conflicting with this marble-moving exercise.[9] Changizi believes that this touch adaption might not be designed for manipulating fine objects but instead for locomotion and supporting body weight when walking barefoot on wet ground or climbing up wet trees or rocks. If we want to find out whether this is the case, the next experiment will need to have a slightly longer risk assessment.

Discriminative touch – the ability to detect our surroundings to a mind-boggling degree of sensitivity – seems almost miraculous, but there are many more layers to our body's most diverse sense. The detection and transmission of mechanical touch has long been understood: mechanoreceptors, which are found at high concentrations in glabrous skin, are connected to the brain by A-beta nerve fibres, which convey signals at the pace of a racing car at full throttle – roughly one hundred and sixty miles an hour. Scientists are just beginning to understand, however, that humans also have another touch system. As well as receptors and their wires geared up for discriminative touch, we have separate ones for emotional touch, which can feel otherworldly.[10] These nerves are called 'C-tactile fibres' and are found in our hairy skin. C-tactile fibres are sensitive to light touch, and their signals travel up to the brain at a much more leisurely pace – roughly two miles an hour.[11] They are not there to give us the facts about what is touching us, but instead carry the emotional signal of the touch. They are optimally triggered by stroking at a speed of 2–10cm a second at an ideal temperature of 32°C, so if you want to strive for the perfect skin-on-skin caress, start with these parameters.[12] These slower signals are processed in areas of the brain more associated with emotion, such as the limbic system.[13] Recent research also suggests that the pleasant activation of this emotional touch-system positively modulates our own sense of body ownership.[14] This has been shown using the 'rubber hand' illusion. (If you happen to be in possession of a lifelike plastic limb, this also makes for an impressive party trick.) Subjects put their right hand on a table in front of them and keep their left hand hidden from view behind a screen, with a rubber hand on the table in its place. Their hidden left hand and the rubber hand in place of their left are then stroked at the same time. When subjects are stroked with the speed and lightness of emotional touch, they are more likely to perceive the rubber arm as their own.[15] Emotional touch enhances our sense of self.

But why does the brush of a hand along your forearm by a lover feel so different to the exact same touch (with the exact same receptors being triggered) given by a doctor during a medical examination, or a brush against a stranger on a crowded train? When someone touches our skin they also touch our brain because these two organs are in a constant dialogue to try to find out what is touching us and how we should react – Who is about to touch us? We're being touched! Is it a friendly touch? Aided by our sight and hearing, our brain starts to form a context for whether what is going to touch us is friend or foe. The expectation of a loving caress temporarily changes the composition of the skin to receive pleasure, while the anticipation of pain makes the physical discomfort on the skin feel worse. When the touch connects, innumerable pieces of information about the pressure, speed and warmth of the touch travel to our brain to add to the context and allow us to interpret it. Expectation, fantasy and reality are intermingled on the surface of our skin, a stage on which our life is played out.

The curious case of our inability to self-tickle, even if we copy the same tactile movements of a teasing friend, gives a fascinating insight into the game of predictions and expectations played between the skin and the brain. We have the remarkable ability to tell the difference between the same sensations caused by our own movements and those from the 'other', whether friendly or a potential threat. This is made possible by the cerebellum, a region of the base of the brain that keeps us balanced and monitors our movements. Professor Sarah-Jayne Blakemore and her team at University College London have discovered that whenever we move our own fingers and limbs, our cerebellum produces an exact mental picture of these movements which then sends a 'shadow signal' to the sensory areas of our brain to dampen down the corresponding sensations on the skin.[16] This helps our skin stay attuned to more important touches by other – potentially predatory – beings.

Blakemore then wondered whether we could trick the brain by using a 'tickle robot', consisting of a piece of rotating foam that strokes the skin by spinning on an adjustable timer. The longer the delay between brushes, the more ticklish it feels, as the robot seems to break the relationship between our brain's predictions of what we should feel and the actual feeling on the skin. The only time that I have come across someone who could self-tickle was a patient with severe schizophrenia. This may well have been down to his brain not fully recognizing that his physical finger movements were his own.

Our inability to self-tickle demonstrates that our brain can exert a subconscious force on our physical feelings that we cannot control. It might be logical, then, to think that our sensory skin is subservient to our brain; that our body is incidental to the power of our mind. To show that this isn't the case, try some of these easy illusions. First, place your left hand in a bowl of icy water and your right in hot (but not scalding) water. After a full two minutes, take them both out and dip them into a third bowl, of tepid, room-temperature water. For a moment our perception of reality is disturbed – one hand thinks that the third bowl of water is cold, the other hot – and it's unsettling. In a similar way, if you simultaneously rub one hand along a piece of smooth plastic and the other across rough carpet and then place both hands on a wall, one hand tells you that the wall is rough, the other that it is smooth. Even though both hands are sensing the same temperature or feeling the same surface, each hand gives the brain a different answer. This shows that the brain gives in to what each hand is telling it. The brain adapts to serve the needs of the skin.

Touch is exquisitely sensitive. It is emotional. It influences our thinking and sense of self. But it is also indescribable. On our journey from sensing to feeling, from the physical to the mystical, we inevitably come to the most compelling powers of our sensual skin: pleasure and pain.

'Sparks fly when we touch' is a common way people respond when asked to describe sexual touch. They use both magical and supernatural words, and say that it is more indefinable than any other human sensation. This often inexpressible feeling has nevertheless provided the ink for a million poems, inspired music and art in every human culture and has started countless wars. Its otherworldliness comes from our mind being found on our skin, where desire and expectation meet physical sensation. In the pleasurable sexual anticipation of skin-to-skin contact, not only do our discriminative and emotional touch-systems kick into action simultaneously, but also our whole skin, our largest sexual organ, changes character. The blood flow to our skin alters to warm the surface, and sweating increases, accompanied by the erection of hairs across the skin – all this further enhances sensitivity to touch; the brain is preparing the skin. On touchdown, the fast mechanoreceptors, the slower, emotional nerve fibres and incredibly sensitive free nerve endings (with which the lips, nipples and genitalia are richly endowed) are all activated.

'Free nerve endings' are the receptors responsible for the detection of many pleasurable and painful stimuli, and are nerve endings that are not attached to specialized cells, such as the four mechanoreceptors we looked at earlier. When it comes to their distribution, our whole body is a unique fingerprint: areas that are highly sensitive and bring great pleasure in some people barely raise an eyebrow in others. Why this variation exists is still unknown, and it may be a long time before we discover this secret of the sexual skin. When these nerves are stimulated, an intoxicating cocktail of hormones – from endorphins (the happy hormones) to oxytocin (the cuddle hormone) – is released. This skin-to-brain conversation crosses from one partner to the other through touch. The physical sensation is heightened by our perception of our partner's reaction to us, as mediated by the skin. As David Linden points out in his book *Touch: The Science of Hand, Heart and Mind*, sexual touch is not just 'a meeting of minds but the meeting of skins'.[17] Our skin delineates 'us' from the outside world, and enjoyable, consensual sexual touch is the ultimate acceptance of the 'other'. On the surface it's easy to think that the skin's role in touch is simply to be the terminus for nerve endings. Sexual touch shows that this could not be more wrong.

Wouldn't it be wonderful to live without pain? Try asking Amjad this seemingly simple question, for this British man of Pakistani heritage has a rare genetic condition, which runs in a small number of families, called 'congenital insensitivity to pain'.[18] Amjad can tell stories of his famous ancestors back in Pakistan who made a living pushing sewing needles and swords through their bleeding skin and walking on hot coals without so much as a wince, in front of captivated audiences. Although their street acts were sold as a demonstration of the power of mind over matter, the truth was that they never felt any physical pain. Most of these street performers did not live to adulthood – one of the reasons why congenital insensitivity to pain is such a rare disease. Amjad, and other patients like him, have described living without pain as 'hell on earth' and 'the devil's curse', as they are forced to spend their life looking over their shoulders, constantly checking their body to make sure that they haven't stepped on glass or burned their hand. They have to learn to replace the protective sensations of pain with a visual reference. Pain, although unpleasant, is a vital warning sign of tissue damage and we cannot live without it. There are also people who acquire insensitivity to skin pain as a result of disease. A leprosy patient I met could not feel with the tips of his fingers, and repeated unnoticed injuries had left him physically deformed. He told me: 'I would rather feel everything than feel the agony of the shame.' Pain is necessary for life, in every sense of the word.

It was originally thought that pain is caused by an excessive activation of the mechanoreceptors in our skin, but it's now clear that skin has specific pain receptors, called nociceptors. This strange word derives from the Latin *nocere*, to harm. Patients with congenital insensitivity to pain are still able to feel other sensations, including fine touch and vibrations. This is because the disease is caused by a mutation in the SCN9A gene, the code for a particular type of sodium channel (a protein responsible for the initiation and transmission of nerve signals) mainly found in nociceptive nerves. Most nociceptors are free nerve endings in our skin that look like the roots of a plant. Each ending is part of a nerve that travels to our spinal cord, where it makes a connection (a synapse) with another nerve, which travels up to the brain.

There are three types of nociceptor: mechanical, thermal and chemical. Mechanical nociceptors detect excess crushing or cutting of the skin. As with pleasure, we have two pain systems, which I found out to my expense. I was walking along a pebbly beach on the Welsh coast. My brother, walking closely behind, managed to catch the back of my flip-flop, causing me to momentarily lose my footing and ram the big toe of my right foot directly into a rock. Immediately a sharp sensation shot up into my brain, causing me to yank my foot backwards. This first wave of pain is carried by fast A-beta fibres and tells us that our skin has been deformed. There was a gap of less than a second before an excruciating, dull wave of pain hit me, causing me to yelp in agony. In a similar way to our emotional touch-system, this pain information is carried by the slower C-fibres. The importance of both of these systems working together is demonstrated by another very rare disease called 'pain asymbolia'. Patients can detect pain – for example, from stubbing their toe on a Welsh rock – but they don't feel any of the unpleasantness. They just feel a strange, almost funny sensation of vibration. The first 'sensory-discriminative' (detecting pressure applied on the toe) pathway is intact, but the 'motivational-effective' (or 'ouch factor') pathway is absent.

Thermal nociceptors detect painfully hot or cold temperatures. The most notable of these is TRPV1. This receptor detects temperatures over 43°C, but it is also triggered by capsaicin, an active component of chilli peppers. When we eat a chilli or it touches our skin, it's not surprising that we feel a burning sensation, as the very same receptors that respond to burning heat are triggered. In a similar way, the main 'cold' receptor on our skin – TRPM8 – is activated by temperatures at 20°C or cooler, but it is also activated by menthol, found in mint. If you dipped a thermometer into toothpaste, or into creams and lotions containing menthol, the reading would be room temperature, but spreading these across your skin provides a noticeable cooling sensation.

Pain detected by our skin is not always a response to external triggers. As a student a week before exam season, I was holed up surrounded by textbooks and pizza boxes. After emerging from a much-needed shower I glanced in the mirror and saw a sudden eruption of little bumps and vesicles (blisters less than 1cm in diameter) on a small strip of skin on the right-hand side of my back, just below the shoulder blade. It was shingles. Whether it was brought on by stress, my revision diet or something else entirely I may never know. But while it was slightly itchy, in my medical-student fascination I was quite excited to see my varicella zoster virus companions finally come out again, having been sleeping in the nerves supplying my skin ever since a bout of chickenpox at the age of three.

Shingles is a visual example of how nerves are distributed in our skin. The area of skin that is supplied by a single nerve coming out from the spine is called a 'dermatome'. There are thirty dermatomes from top to toe, and the invisible boundaries of these strips of skin are only revealed by shingles. After an outbreak of chickenpox, which often affects large swathes of skin, the causative virus, varicella zoster, retreats to a single nerve root in the spine. Here it can lie dormant for years. For reasons poorly understood, but likely related to a weakened immune system, the virus can travel up the nerve from the spine to the skin, causing an eruption limited to one discrete area of skin on one side of the body. The blistering and redness of the shingles did not cause much of a nuisance, but roughly a week after the visible symptoms had died down I developed a complication called post-herpetic neuralgia. Sharp, burning pain tormented that area of my back for the remainder of the exam period and beyond. It was impossible to put any pressure on the area, making sleep very difficult, which compounded the misery of trying to forget the pain during the day. This is an example of 'neuropathic' pain, often experienced on the skin. As opposed to our specialized nociceptors detecting pain stimuli, neuropathic pain is caused by damage to our nerve endings sending pain signals up to our brain.

But nociception – the detection of harm – is not synonymous with pain. Pain is a phenomenon, a painting drawn in our mind by physical nerve impulses, our emotions and our mental state. The multi-dimensional subject of pain is extraordinarily complex but can be condensed into a brief, if rather simplistic, analogy. If our conscious mind – where we finally register the physical and emotional feeling of pain – were a castle we would have a series of fortified gates controlling the messengers of pain entering the fortress. Most of these gates in our skin open after reaching a stimulation threshold, whether it is mechanical, thermal or chemical. The bricks that make these gates, and consequently their relative strengths to different stimuli, vary by sex, genetics and culture. At my school, during the inescapable playground fights, the person to avoid was Duncan, a large, red-haired Scot. As well as being a good foot taller than the rest of us, he claimed that his terrifying strength lay in the fact that 'Scottish people don't feel pain!' He certainly packed a powerful punch. Fascinatingly, recent studies have shown that red-haired people are indeed more resistant to a number of types of pain, including electrical pain, but are more sensitive to thermal pain.[19] This could well be due to little-explored effects of a mutation in the melanocortin-1 receptor (MCR-1) gene, which produces the red hair pigment.

Continuing our castle metaphor, we can to some extent open or close the gates ourselves. Physical stimulation of non-pain receptors (such as receptors that detect vibration) can actually reduce pain. This is why rubbing the skin over a freshly knocked knee can dampen down the pain, at least temporarily. In many cases control over these gates is not physical, but mental. As a doctor, I find taking blood from patients and giving injections as much second nature as brushing my teeth. When it comes to visiting the doctor myself, however, my fear of needles has not eased since childhood. The anticipation starts on entering the surgery, and builds in the waiting room, and even the neutral tones of 'Dr Lyman to Room Three, please' gets my pulse and mind racing. This concoction of overthinking and emotion makes a small pin-prick, which most people barely notice, feel as though I'm being skewered by a jouster's lance. Why I remain so terrified is another question; it could be the fear that the doctor or nurse giving the injection may just happen to be one of my more incompetent friends from medical school.

The opposite end of this spectrum can be seen in what has been called 'the most British conversation ever'. At the Battle of Waterloo in 1815, Lord Uxbridge, a British aristocrat and officer, was riding alongside the Duke of Wellington. Lord Uxbridge had just led a number of cavalry charges at the French, with cannon-balls whizzing overhead, blasting into his soldiers either side of him and shooting down eight horses from underneath him. Exhausted, full to the hilt with adrenaline and utterly focused on the task in hand, it took him a while to notice that one French cannonball had completely shattered his right leg. His following words should be read in as smart an English accent as possible: 'By God, sir, I've lost my leg'; to which Wellington replied, 'By God, sir, so you have!'[20] In the many conversations I've had with soldiers in one of the world's leading military hospitals in Birmingham, they report that even the most devastating injuries can feel painless in the heat of battle. The ancient Roman philosopher Lucretius recorded how when 'the scythed chariots, reeking with indiscriminate slaughter, suddenly chop off the limbs' of men, the 'eagerness of the man's mind' means that 'he cannot feel the pain' and 'plunges afresh into the fray and the slaughter'.[21]

Have you ever wondered why you jerk your hand away from a searing-hot plate seconds before you actually feel the burning pain across your fingers? Your body reacts before you can think, your skin seeming to achieve time travel. Skin receptors detect the heat of the plate and fire nerve impulses from your fingers, up your arm and to your spinal cord via a sensory neuron. Within the spinal cord, the impulse is passed on to a motor neuron via a small relay neuron. The impulse from the motor neuron then travels to a muscle and activates it, yanking your hand away from danger. None of this occurs in the brain, making it truly unconscious. Pain is then recognized about a second later as the impulses from the slow pain nerves, the C-tactile fibres, reach the brain.

It is a curious fact that feelings of pleasure are only fleeting, yet too often pain persists. Pain can leave its mark on our skin, sometimes for a lifetime. When we damage our skin, it changes for a time to let us know not to damage it again; if we get sunburn, even the most casual of touches can feel like a stinging slap and a warm shower becomes a searing torrent of lava. This effect, in which usually innocuous sensations bring about the feeling of pain, is called allodynia. The damage from the initial lesion (whether it be sunburn or a splinter) brings about an inflammatory cocktail of molecules, including proteins called cytokines and lipids called prostaglandins, which lower the threshold of our pain receptors, making the nerve endings in our skin particularly touchy for a time. As we've seen before, it's a two-way conversation across our nerves, and in response to pain our nerve endings, too, can release inflammatory molecules, further lowering the pain threshold across the skin. Our whole skin changes in response to pain, encouraging us to protect the damaged tissue, and teaches us a lesson.

But this does not explain the chronic pain in our skin that continues months, even years, after all traces of physical damage have gone. Stimulation and damage to nerve endings in the skin can have long-term effects on the other end of the nerve, which lives in our spinal cord. Changes in signalling between nerve connections, and indeed the growth of new connections, can make permanent 'pain memories' in the spine. These then constantly relay pain signals into the brain, even when the damaged skin has been fully repaired. New research suggests that pain and nerve damage can create 'epigenetic' changes in our nervous system, where the make-up of the cells is changed for the rest of our lives, leaving footprints of the original pain.[22] Fascinatingly, the change in synaptic communication is similar to how new memories are formed in the brain. This is in addition to the cognitive and emotional memories we form after experiencing pain.

On a trip to a hospital in a remote corner of India, I came across a patient whose mind retained the ability to experience pain on his skin even though the skin was not there at all. Ten years earlier, Aman had been driving his brightly coloured truck on the ten-hour journey from the plains of Northeast India up into the jungle-clad foothills of the Himalayas, somewhere near the Burmese border. The steep, winding roads carved into the mountainside were mud tracks even on a good day, and it is a terrifying, vertiginous journey I wish never to repeat. Aman was trying to make the ascent in a full-blown monsoon. About halfway up the ascent, the mountainside above him suddenly collapsed in a slump of mud and rock, slamming into his orange truck and sending him careering off the mountainside. The vehicle fortuitously smashed against a clump of thick trees sticking out of the cliff a few metres below, otherwise he would have fallen to a certain death. Aman's right arm took the full force of the landing, however, with his forearm and elbow completely crushed. After a lengthy rescue from this precipitous position, the local doctors decided to carry out a successful below-shoulder amputation in the hospital. As Aman and I talked, his face intermittently screwed up into a fleeting grimace. He explained that several times each day he got the sensation that his whole right arm was still there, accompanied by the feeling that his invisible fingers were being scalded with boiling water. It appears that over half of patients with amputations experience a similar type of 'phantom limb pain'. I had long assumed that the cause of this bizarre condition was damaged nerve endings in the stump sending aberrant pain signals up to the brain. The relevant literature, however, explains that even when further amputations of a stump have been conducted in an effort to remove the remaining nerve endings, the pain does not stop. In fact, it is often made worse. Interestingly, surgeons have discovered that when a local anaesthetic is applied around the area to be amputated (as well as the patient being administered a general anaesthetic), the incidence of phantom limb pain is dramatically reduced. This suggests that the body forms 'pain memories' during the amputation in a similar way to the formation of memories in the brain. For Aman, this memory took the form of invisible, burning skin.

There is, however, a sensation on our skin that many say is worse than pain. In the Bible, when Satan has to choose a physical punishment to torment the God-fearing Job into atheism, he picks the simple itch. In Dante's *Inferno*, the worst sinners who occupied the eighth circle of hell were subjected to 'a fierce itching that nothing could relieve'. The itch can be the most gentle, but also most malevolent, caress. On a trip to the Libyan desert in North Africa, a Second World War historian told me that the itch from the minutely small legs of the region's flies had driven many soldiers to what the French call *le cafard*, or desert madness. The unpredictable, unreachable and unceasing itching from these airborne hordes even drove one British soldier to try to shoot the flies with his revolver.

But unwelcome outsiders are not the only cause of itching. Itching can sometimes come from *inside* the skin. I recall a patient who almost lost her right foot to infection after scratching through the skin of her ankle in a futile attempt to relieve the irritation of itching. Internal itching can be caused by a number of diseases, including iron deficiency, anaemia and liver disease. One of the strangest itches is known as acquagenic pruritus: a mysterious, intense itching following the skin's contact with water.

There are many causes of this frustrating feeling. Perhaps the most well known is the molecule histamine, released from the mast cells in our skin during inflammation, which causes allergic rashes or the itch of a mosquito bite. What makes itching so powerful is its urgency. Its use in language and culture – 'itching for a fight', 'the seven-year itch' – shows how this skin-specific feeling is the perfect physical manifestation of an irrepressible urge. Unsurprisingly, its response, the scratch, is associated with extreme pleasure, accompanied by feelings of both sin and guilt. The French philosopher Michel de Montaigne noted that: 'Scratching is one of the sweetest gratifications of nature, and as ready at hand as any, but repentance follows too annoyingly close at its heels.'[23]

Itching has traditionally been considered to be a weak variety of pain. It is easy to see why. Both cause discomfort and bring about an immediate protective response (pain withdraws a hand from a boiling plate, an itch makes us scratch away a poisonous scorpion or a disease-ridden fly) and both are mediated by cognition and emotions. This notion was turned on its head in 1987, however, when German scientist H. Handwerker discovered a curious difference between these feelings.[24] If an itch were considered to be 'weak pain', it would follow that an increasing quantity of 'itch' would lead to actual pain. However, when his team inserted increasing doses of histamine into the skin of human subjects, they became progressively more itchy but didn't feel any pain. Itching is now known to be a completely separate system from pain, with information travelling to the brain along entirely different pathways. An itch nerve fibre can detect sensations over a large area of skin (centimetres in contrast to the millimetres of pain fibres) and the nerve impulses are much slower, which is why an itch gradually waxes and wanes.

A recent study has also found that a molecule called brain natriuretic peptide (BNP) transmits itching sensations from our skin to our brain without triggering any pain sensations, potentially paving the way for new anti-itching treatments.[25] A particularly intriguing aspect of the difference between these unpleasant feelings is that, for most people, thinking about burning our hand on the stove or watching a violent Hollywood war movie does not induce feelings of pain, although the mere mention of lice can get us scratching.[26] When a German professor gave a lecture in which the first slides consisted of images of bugs and people scratching, and the second half showed 'soft' images of babies' skin, a hidden video camera revealed that the students subconsciously scratched far more in the first half than the second.[27] Your skin may even be crawling as you read this.

It isn't understood why we tend to scratch ourselves when we see pictures of insects or see others itching or scratching. One theory is that the response is designed to remove skin parasites that might be making their way around a community. Socially contagious itching was originally thought to be based on humans showing empathy to other members of the group, resulting in itching that would reduce the chance of observers being infected by parasites, but it may be more impulsive than that. In 2017 Dr Zhou-Feng Chen of the Washington University School of Medicine in St Louis found that mice placed in a cage next to mice with a chronic itch were likely themselves to start scratching.[28] Socially contagious itching may in fact be hardwired into our brain; when mice saw their friends scratch, their brains immediately released a molecule (called a gastrin-releasing peptide) causing them to scratch as well. If this molecule is blocked, the mice no longer socially scratch, but still scratch themselves if they are exposed to an itch-inducing stimulus, such as histamine. These separate itching mechanisms could provide clues as to why we have other socially spreadable behaviours, such as yawning.

The complex worlds of pain and itch show us that the skin and the mind communicate over millions of individual pathways, starting from receptors, travelling along nerves and into various uncharted regions of the brain. Many of these individual journeys are pushed in different directions by the tides of emotion, memory and cognition. The physical distance between the skin and the brain, however, is also a philosophical one. We like to think that we see and feel the world directly, but much of the picture of the world our mind constructs is in fact a necessary illusion. Phantom itches, which we all experience, seem physically real, but are constructs of the mind. In the case of Aman and his phantom skin, his brain held an imprint of a limb that no longer existed. In our other senses, too, we can see why we need illusions to deal with reality. If we see someone clap, we perceive the sight and sound at the same time. The brain, however, is processing these two discrete pieces of information travelling at different speeds and as a result our mind sees what happens in the world roughly half a second after it actually happens. This is because only 20 per cent of the fibres that reach the visual areas of the brain originate in the eye; the rest come from memory areas of the brain. Our reality comes from the image of the world we construct in our mind via our senses, with the brain unconsciously filling in the gaps in the limited signals we receive. The vital reception of signals in our skin is a bridge – albeit sometimes a very long one – between the physical reality of the outside world and our perception of it: the world we create in our head. Our skin is truly an extension of our minds.

Touch is a most extraordinary sense, enabling our skin to be a sensitive instrument detecting and protecting our journey through life. But when skin touches skin, there is a seemingly mysterious, almost magical, transfer of power. In the 1960s Dr Sidney Jourard embarked on research most academics could only dream of. The Canadian psychologist travelled around the world (to conveniently fashionable locations) for a spot of people-watching. He would sit in the corner of a café and count the number of interpersonal touches made by the locals in one hour. Puerto Rico came top with 180 touches an hour, Paris fared well with 110, but London – my hometown – came in at a pitiable zero touches an hour, living up to our uptight, touch-averse stereotype.[29] Although we rarely think about the effects of a handshake or a pat on the back, studies are revealing that everyday tactile sensations can profoundly influence our social judgements.

Imagine that you are at a small corner table in the fictional Café Touché in Paris. In the manner of Dr Jourard, you are keeping an eye out for any fleeting touches. The effects of all the touches you will see have been demonstrated in psychological studies.[30] On the table to your left, two lovers are debating whether to book an expensive holiday. At the point at which the man's finger is hesitantly hovering over the 'pay' button on his phone's browser, the woman reaches over and reassuringly holds the back of his other hand. He books the holiday, and they leave the café holding hands. A woman's touch can prompt a man to take more risks, but intriguingly not the other way round.[31] Holding hands is also a social 'tie-sign' indicating the partners' exclusive bond. To your right a waitress is engaged in animated conversation with a customer and playfully touches his arm as she places the bill on the table. This fleeting, socially subconscious touch makes the patron likely to increase his tip, by as much as 20 per cent. In the opposite corner, away from the window, a nervous young chef is being interviewed by the head chef for a job in the restaurant. The head chef is holding a heavy clipboard in front of her. Experiments have shown that an interviewer handling a heavy clipboard or folder, rather than a light one, is more likely to hire someone.[32] The suggestion is that our touch's perception of physical weightiness can bias our thoughts regarding another person's intellectual or practical heft; the sense of solidity is transferred on to them.

On a small table next to the door, a sales representative is sitting opposite her client. Although this is their first face-to-face meeting, the cup of coffee in the client's hand warms her to the sales rep and the soft cushion on which she sits makes her more likely to agree to the deal. As they get up to leave, they briefly shake hands and the sales rep reassuringly touches the client on the forearm. They depart, smiling – by deft, unconscious and unintentional uses of touch, the sales rep has ensured she will be allowed a follow-up meeting. A man now enters the café and is quickly smothered in a noisy and exuberant embrace from an old college friend. The hug releases a potent mix of 'happiness molecules', including oxytocin and endorphins, strengthening and affirming the bond between the two. Behind the bar, the aproned baristas and waiters frantically run after orders and, unable to find time to speak, encourage each other with the occasional pat on the back and teasing nudge – these actively help the bar team to bond and improve their work environment. Studies of basketball teams show that the teams with more physical touches on court (whether it be high fives or fist-bumps) are more likely to be successful than those with fewer touches.[33] It would be interesting to see whether these regular touches have a similar effect on fist-bumping tennis doubles partners.

One study carried out at the University of California, Berkeley, separated two strangers from each other by a thin wall.[34] One participant put their arm through a hole in the wall and the other person had to convey an emotion by means of just one second of touch. Remarkably, the receptive participants were largely able to differentiate between feelings of compassion, gratitude, love, anger, fear and disgust from these fleeting touches.

Touch doesn't just communicate; it also heals. In the early 1200s, the Holy Roman Emperor Frederick II carried out an experiment that would certainly not be granted ethical approval today. He set out to discover the original human language. His trial design was to separate babies from their mothers at birth and keep them in conditions where the nurses-cum-researchers were banned from speaking in their presence and could not even touch the infants. Salimbene di Adam, the medieval Italian chronicler who first recorded this story, observed that Frederick never got to hear the babies utter their first words because 'the infants could not live without touching'.[35] Even though they were fed, they died before they could speak. The findings of this grotesque experiment have been repeated throughout history. There are thousands of Romanians alive today who bear the scars of touch deprivation. Raised in critically understaffed Romanian orphanages during Nicolae Ceausescu's drive to increase the country's population in the second half of the twentieth century, they have much higher rates of both physical and mental illness – from diabetes to schizophrenia – than the rest of the population.[36] While other factors, such as deprivation of verbal communication, play a part in the Romanian story, it's clear that physical touch is necessary for good physical and emotional health. It is a language of love and compassion, and it is critical for human development.[37]

It is sobering that much of our knowledge of the role of touch in human development and survival has come from such crises of care. In 1978 the neonatal intensive care unit of the Instituto Materno Infantil in Bogotá, Colombia was struggling from being understaffed and from a lack of incubator space; most worryingly, it had a mortality rate of about 70 per cent. Dr Edgar Rey Sanabria decided to try something radical. He asked for preterm babies to be held skin-to-skin to the chests of their mothers, for warmth (effectively replacing the incubators) and to encourage breast-feeding. The mortality rate suddenly and unexpectedly dropped to 10 per cent.[38] It was quickly apparent that skin-to-skin contact with their mothers was a remarkable healing agent for these babies. This 'kangaroo care' became increasingly popular across the globe over the following decades, and a growing number of studies are confirming that the skin of the mother or caregiver has a remarkable power.[39] A 2016 review found that kangaroo care improves vital signs (such as heart and respiratory rates), helps with sleep and increases weight gain.[40] Another study found that in the developing world infants were 51 per cent less likely to die in the first month after birth if given kangaroo care in the first week of life.[41] Skin-to-skin contact also goes both ways: it's psychologically beneficial for parents (dads benefit from skin-to-skin contact with their babies, too), reducing anxiety and increasing confidence in their parenting skills.

The healing properties of touch are not restricted to preterm infants. I remember, as a medical student, sitting in on consultations with a family physician, sceptically wondering at the non-clinical nature of her phrase 'love travels through the skin' as she advocated giving a reassuring hand-hold or a friendly pat on the back to some of her patients. However, soon afterwards I read a study that found that patients placed in an MRI scanner and told that they were to receive electric shocks experienced dramatically reduced levels of stress when a romantic partner held their hand.[42] Couples in another study, who were shown how to touch each other in emotionally sensitive ways over a longer period of time, experienced less stress and had lower blood pressure compared to a control group.[43] Research continues to demonstrate that skin-to-skin touching and physical hugging stimulate nerves, releases endorphins and oxytocin, and activates reward and compassion centres in the brain. The short-term happiness that results may not cure an infection or prevent cancer, but it does reduce stress and improve psychological wellbeing, both of which can also strengthen our immune system. It's not all about short-term chemical changes, though. Motherly touch, as shown by grooming animals, creates lasting 'epigenetic' changes in their offspring; fingerprints of their care that stay with the children their whole life, improving health and reducing stress. At the other end of the age spectrum, it has been discovered that touching people with Alzheimer's disease helps them make better emotional connections with others and dampens the symptoms of this devastating disease.

As a teenager I spent a considerable amount of time staring at the black line at the bottom of a swimming pool, running across muddy fields and battling against the English rain on a bicycle in an effort to join the British triathlon team. That also meant many hundreds of hours in sports massage. The benefits of muscle manipulation seemed obvious, but I never wondered whether the meeting of skin with skin also had a positive effect on my wellbeing. Professor Tiffany Field's team at the University of Miami has discovered many health benefits to massage.[44] Giving an elderly patient a massage as part of a social visit improves their cognitive and emotional functions far more than a social visit alone. A massage intentionally given with emotion has more beneficial effects than one without, which in turn is better than the same movements given by a massage chair. Massage has also been shown to be hugely calming to many people with autism, where previously it was assumed that those with the condition are averse to all forms of human contact.

It has been known for millennia that the laying-on of hands imparts healing power, but we are only just beginning to understand how this actually works. Touch has a potent emotional quality that both biologically and cognitively makes us feel loved and relaxed, and which in turn reduces stress. This easing of our brain–body conversation manifests itself in many physical ways, from reduced blood pressure to improved immunity. The healing power of touch truly does affect our mind and our hearts and, as research progresses, the power of human touch is sure to throw up some more surprises.

The sensational abilities (literally and figuratively) of skin have also led to the development of civilization and to human dominance over nature. Technology (from the Ancient Greek *techne* and *logía*) can be loosely translated as 'the study of handcraft'. We have been able to create and control information through our fingers, enabling us to craft societies – from manipulating tools to carving hieroglyphs, from touch-typing at one hundred words a minute to operating the magic mirror that is the touchscreen of a smartphone. Whereas 'resistive' touchscreens detect a bend in the glass when you press it and send an electrical signal to the appliance's computer, more recent 'capacitive' smartphone touchscreens harness a lesser-known quality of our skin. Directly underneath the glass of capacitive touchscreens is something that resembles the street plan of New York City. Running top to bottom are minutely thin wires of conductive metal called 'driving lines' that supply a constant flow of electric current. Running left to right are 'sensing lines' that detect the current. Whenever our fingers touch the screen, they attract the current, creating a voltage drop. The electrostatic fields created by the screen's criss-crossing wires are distorted and pass on incredibly detailed information to the phone's computer about the location, power and – if swiping – direction of the touch.[45] Our fingertips may not have the thunderbolt-generating power of Zeus, but human skin is an electrically conductive material. No amount of pressure applied by non-conductive materials can trigger a conductive touchscreen, which is why it doesn't work when we wear gloves. The next time you're scrolling through social media, wonder at the fact that your skin is part of the electronics.

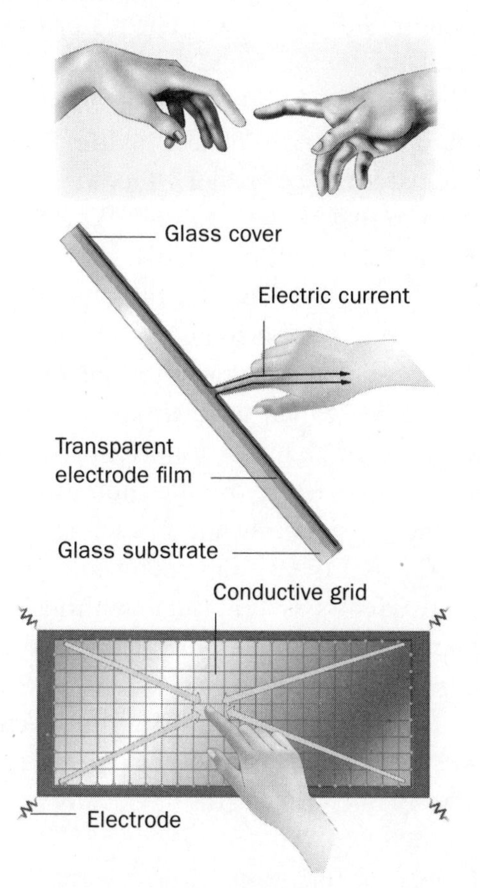

ADAM'S TOUCHSCREEN

Conversely, the incredibly sensitive apparatus in our skin can also connect with technology to receive messages. A remarkable Frenchman discovered this in the early 1800s. Louis Braille's father was the best harness-maker east of Paris and, as a child, Louis wanted nothing more than to follow his father's trade. He would waddle around the leather workshop and try to imitate his father's skilled hands. One morning, when Louis was three years old, his father stepped out of the shop to talk to a customer for a few minutes. Louis picked up a sharp metal awl and tried to punch a hole through a sheet of leather. The awl slipped from his hands and instead penetrated his left eye. The resulting infection spread to his other eye and, by the time he was five, Louis was permanently blind. Unlike most blind people at the time, who had to go out and beg in order to try to make a living, Louis was fortunate enough to have parents who fashioned for him a wooden cane and encouraged him to feel his way around. When he was ten, Louis attended a school for the blind in Paris, set up by Valentin Haüy. There were few books available for blind people, and those that existed used Haüy's cumbersome system in which embossed pieces of lead were shaped into letters of the alphabet on the pages of large books, which were impossibly heavy. The delay between the speed of Braille's mind and the speed of reading Haüy's script was unbearable. Braille then came across the work of Charles Barbier, a captain in the French army who had developed 'night writing', a twelve-dot secret military code. While an improvement on Haüy's system, night writing was still cumbersome and fiendishly difficult to understand. Braille's genius lay in his simplification of this code into six raised dots in two columns, enabling the recognition of a letter with the touch of one finger.[46] With this basic but ingenious technology, Louis Braille invented a system that has enabled thousands of visually impaired people to read through touch.

A modern leap in the study of how we can receive information through touch is in the area of haptic technology, where information is transferred to the user in the form of vibrations and movements. As a teenager, I remember the buzz of the crude vibrations in my videogame controller when my race car veered off the track or I was hit by enemy fire. Technology has advanced dramatically since then, and touch communication is now the final frontier of virtual reality (VR). With a VR headset it is easy to mimic visual and auditory stimuli, but full immersion is impossible without the sense of touch. 'The lack of touch breaks the suspension of disbelief in VR,' Katherine Kuchenbecker from the University of Pennsylvania has observed.[47] Her team has helped develop a digital thimble controller that recreates what it is like to touch a range of objects by vibrating at an extraordinary variety of frequencies.[48] It also detects the position of the fingers in space and calculates a wave-force (called a dynamic tactile wave) that makes one feel virtual objects differently when the finger changes direction. Blending vibrations with movements and other visual and auditory sensory inputs fools the brain into thinking we are holding something physical when in actuality we are merely grasping the air. The possibilities of this technology range from being able to feel the fabric of a dress at home before we buy online, to helping trainee surgeons 'feel' the tug and pull of equipment on human organs before they are allowed to operate on real patients.

Huge bounds are being made towards making robots more human-like. Robots can build cars and perform surgery with incredible precision. They can hold a conversation and even develop their own languages. Robots can out-think us at chess and outperform doctors in diagnosing disease. A novel by a robot was even shortlisted for a Japanese literary prize. There is considerable research into mimicking human touch in robots, with particular hope for the development of lifelike prosthetic limbs. I recently enjoyed interrogating one of my friends, who works in a robotic institute, with questions on the future of robotics and how it could shape our world. After the usual 'Will robots take my job?' and

'Will robots take over the world?' (neither of which he had ever been asked before, I'm sure), I asked whether we could build robots with surfaces replicating human skin, and thus simulate the intricacies of human touch. His answer was telling: 'The trouble is that when it comes to touch, we just think of the skin as a terminus for nerve endings. Getting a robot to pick up a set of keys is hard enough, but getting them to *feel* is something else altogether.' It is conceivable that one day we could develop a bionic skin for robots that exactly mimics the power of our 'fantastic four' mechanoreceptors, from slip detection to force control.[49, 50] Indeed, a stretchable skin for robots was developed in 2017 that can start to detect crude changes in shear forces and vibrations.[51] But the power to convey, transmit and receive emotion across the skin, and how it combines the physical and the social in an inordinate number of complex ways, currently seems beyond the limits of engineering. Maybe touch is what makes skin our most human organ.

Recent developments in touch technology and robotic touch are laced with irony; as a society we are in danger of 'losing touch'. We're more comfortable with the interaction between our finger and smartphone screen than with a comforting hug or reassuring pat on the back. Our most ancient sense is mystical, sometimes literally indescribable, and we mustn't forget its power in emotional communication, social bonding, health and survival. The sense of touch makes skin simultaneously physical, emotional and transcendental. The Italian artist who painted the supernatural touch on the ceiling of the Sistine Chapel understood that, as he said himself, 'to touch can be to give life'.

7

Psychological Skin

How the mind and skin shape each other

'Strategies of concealment ramify, and self-examination is endless.'

JOHN UPDIKE, *SELF-CONSCIOUSNESS*. THIS CHAPTER 'AT WAR WITH MY SKIN' WAS DEDICATED TO HIS PHYSICAL, MENTAL AND SOCIAL BATTLE WITH PSORIASIS).

THE SQUAT MAASAI hut perched on the edge of the village, itself alone in an ocean of grassland bordering the Serengeti. I sat cross-legged on the floor directly opposite our host, Remi, who had invited me and Albert, a local doctor, to make use of his encyclopaedic knowledge of Maasai herbal medicine. We discussed the medicinal uses of savannah flora for humans and, more importantly for the locals, cattle.

After we had talked for a while, Remi beckoned into the hut a fourteen-year-old boy who had what his village and family considered to be an 'incurable skin disease'. The boy was exhibiting a lumpy, violet-coloured rash with blistering across his forehead and both cheeks. The remainder of his body was unaffected. Tense blisters around the eyelids caused him to wince each time he opened his eyes. The condition had erupted some months ago and had been worsening. When Albert asked me to offer a diagnosis I was bemused by the odd pattern on the boy's face. I glanced at Albert, who also looked flummoxed. A few probing questions, however, revealed that the boy was due soon to be initiated as a *moran*, a warrior. This entailed a test comprising months of walkabout, far from home. It was apparently not much more pleasant than the old practice of having to spear a lion. As the boy's history began to unravel, it slowly became clear to us that he was secretly rubbing his face with the leaf of a specific savannah plant whose chemicals were widely known to cause sun-sensitivity and blistering. He was intentionally and successfully mimicking – indeed, creating – a physical skin disease so that he could stay at home to avoid the rite of passage. This condition, where the patient intentionally causes skin damage to simulate disease, is called 'dermatitis artefacta'. It is a physical manifestation of a psychological problem. As well as feigning illness, dermatitis artefacta can bubble up from many springs of our psyche: from a call for attention following abuse or trauma, to the desire for medical attention in Munchausen syndrome.[1] Albert jokingly called the boy's condition 'Monday syndrome' as it often presents in children trying to avoid school. Skin diseases, particularly visible ones, are not simply physical but also psychological.

Skin is the cape of a continent, where we can look out at the blurry boundary where the oceans of mind and body meet. The relatively new field of psycho-dermatology navigates this border between the visible and invisible.[2] Interestingly, the brain and the skin develop from the same layer of cells in the embryo – the ectoderm – and these old friends seem to be reunited at various points throughout our lives. The dynamic relationship between skin and mind, once an area of mystery and scepticism, is constantly being confirmed by science.

Skin–mind interactions are exceedingly common but remarkably overlooked. They can be split into three groups, with the caveat that they are not entirely mutually exclusive:

1. THE MIND TOUCHING THE SKIN: How our mental state can affect the physical state of our skin, such as psychological stress exacerbating psoriasis.

2. THE SKIN TOUCHING THE MIND: Living with a visible skin condition can have various emotional and psychological effects, such as the depression that so often accompanies acne.

3. PSYCHIATRIC CONDITIONS THAT MANIFEST THEMSELVES IN THE SKIN: From obsessive skin-picking (dermatillomania) to the Maasai boy's dermatitis artefacta, these conditions are more uncommon than 1 and 2, and often more unusual, but they can be devastating.

Imagine you, too, are sitting alongside the group of cheerful village elders. There's an air of relief now that the mystery of the blistered boy has been solved. The long trip back to town awaits us so we bid our farewells and head off towards the cars. As we leave the enclosure you take in the view. The sun starts to dip behind the horizon, bathing the Serengeti in a golden glow. The flat-topped acacias cast long and gentle shadows. You have to take that last photo; you break off from the group to get the perfect vantage point. Five minutes later you're alone on a dusty hillock, peering through the camera lens.

Something in the foreground catches your eye.

No more than fifty metres away, thinly veiled behind a tuft of savannah grass, is a lone, muscular lioness. She is watching you.

Your entire being sharpens. Briefly stuck in suspended animation, you notice your chest pounding and lungs expanding. You're acutely aware of the existence of every muscle in your body, ready for swift, violent action. All that matters now is whether to flee or to fight. Your bladder and bowels prepare to empty. Your heart pumps harder and faster to squeeze oxygen into your muscles and ready them for battle. Blood drains away from your face and seems to be replaced by copious amounts of sweat. Hairs stand on end, poised, each follicle seeming to mimic what is happening to your entire body.

This is what stress looks and feels like, and stress is of course incredibly important. The 'fight-or-flight' response, which is caused by a largely unconscious firing of specific nerves (collectively called the sympathetic nervous system), makes us momentarily superhuman and it has been integral to human survival for millennia. Statistics suggest that you probably haven't been in a life-or-death stand-off with a lion, but you'll recall having some of these symptoms minutes before an important job interview or speaking in public. The role of the skin is important not only in the seconds or minutes of this response – with the sweat cooling down the body for the fight and the face draining of blood to supply the muscles – but also in the following hours or days. In this period after a fight-or-flight response the skin's whole immune make-up changes. This psychological stress increases inflammation in the skin for days; this could be to prepare it to fight the infections that come with a lion's bite.[3] But this fight-or-flight response certainly isn't the only way that psychological stress affects the skin.

During periods of stress, a tiny part of our brain called the hypothalamus secretes corticotropin-releasing hormone (CRH). CRH stimulates the pituitary gland, which is also inside the skull, to secrete adrenocorticotropic hormone (ACTH). ACTH travels to the adrenal glands, which sit on top of the kidneys, and induces them to produce cortisol. Cortisol and CRH have potent effects on skin inflammation but, confusingly, they can each both increase and decrease inflammation in certain contexts. Cortisol enhances immunity and inflammation, but at high doses it can dampen inflammation. A common example can be found in steroid creams that hold back the inflammation of eczema. These creams contain high levels of cortisol and are designed to decrease the body's natural immune response. Another path mental stress can take to provoke the skin is often referred to as neurogenic inflammation (inflammation arising from the nervous system). Nerve endings in the skin contain a number of inflammatory substances, the most well known being 'substance P'.[4] The nerve endings release the substances during stress and contribute to the mayhem. Hormones

Check Out Receipt

Livingston Public Library
973-992-4600

Tuesday, November 17, 2020 11:29:29 AM

Item: 39106091777839
Title: The Kidnapping Club : Wall Street, slavery, and resistance on the eve of the Civil War
Due: 12/01/2020

Item: 31792006045954
Title: The remarkable life of the skin : an intimate journey across our largest organ
Due: 12/01/2020

Total items: 2

You just saved $60.00 by using your library. You have saved $968.94 this past year and $1,533.94 since you began using the library!

Thank you for visiting the
Livingston Public Library!

livingstonlibrary.org
(973) 992-4600

such as adrenaline and CRH, as well as neurotransmitter molecules such as substance P, cause 'mast cells' (the skin's equivalent of land mines) to release potent inflammatory molecules. The molecules increase the diameter and permeability of skin blood vessels to allow cells from the body's immune system to come to the scene as quickly as possible, but they also irritate nerve endings, which causes itching and encourages the further release of inflammatory substances. And down we go in a spiral of inflammatory despair.

Mental stress even changes the character of the skin's immune system. 'T helper cells' are critical immune cells in the skin and can be divided into a number of subtypes, each with different 'personalities'. Humans usually have a healthy balance of T helper 1 (T_H1) cells, which tend to fight viruses and bacteria that live inside cells, and T helper 2 (T_H2) cells, which focus more on attacking bacteria and parasites that live outside cells. Mental stress tilts this balance towards a T_H2 environment, which can cause the red, itchy inflammation seen in eczema.[5] Even the moderate stress caused by repeatedly-ringing mobile phones has been shown to completely modify the skin's immune personality. In acute (short-term) psychological stress, the immune response is increased; not only are immune cells that live in the skin (such as mast cells) activated, other immune soldiers are recruited to the skin from other parts of the body through the blood. Inflammation prepares for the infection from the lioness bite and also has an 'adjuvant' effect. This means that stress stimulates our immune system so that it is better prepared to recognize new foreign microbes about to pass into the body from breaks in the skin. A study in 2017 also found that stem cells in the skin can actually 'remember' inflammation, closing and resolving future wounds in the same location faster.[6] This short-term burst of inflammation tries to protect us, but for those with a pre-existing skin condition it is responsible for the brief but undesirable flare-up of psoriasis or the eruption of acne.

Chronic (long-term) stress lasts from days to months – sometimes longer – and it is a very different beast. Chronic stress can both increase inflammation and reduce the immune response, both of which are bad. In essence, it unbalances the skin and can cause cycles of worsening disease. In eczema, long-term stress skews the T_H1–T_H2 balance in favour of T_H2, which worsens the disease. Chronic stress also accelerates the ageing of skin and its appendages. A series of images of Barack Obama's hair across his eight years of presidency could pass for shades of grey in a paint catalogue. The weight of the presidency is also evident in the lines and wrinkles across the skin of his face, appearing at a much faster rate than with normal chronological ageing. Dutch photographer Claire Felicie witnessed this process over the course of just one year, capturing shots of Dutch Marines in Afghanistan before, during and after deployment. In these powerful photos the subtle but significant effects of stress-ageing are all too apparent.[7]

Psoriasis has one of the strongest associations with stress, with surveys of Americans and Europeans finding that emotional stress is the number-one trigger for flare-ups.[8] It is no surprise that the 2008 financial crisis saw a record peak in psoriasis and eczema consultations.[9] For too many people psoriasis is a vicious circle, sometimes spiralling out of control. The appearance of the psoriatic plaques causes physical discomfort and social anxiety, the stress of which worsens the condition, at which point it often spreads to more visible areas of the skin, resulting in even more stress, social exclusion, depression and anxiety.

There is also strong evidence that long-term stress wears down and blocks the skin's garrison of immune cells, something called 'immunosuppression'. You may have had experiences of stress seeming to cause cold sores around the mouth or the appearance of shingles on the skin. This is because the devious viruses that cause these conditions, herpes simplex and varicella zoster, respectively, belong to the herpes family. These pathogens are unique in that they exhibit 'latency', which means that once they have infected a human host, they silently stay with them, nestling in nerve endings throughout the rest of their life, awaiting reactivation. One theory as to why they reappear is that long-term stress suppresses the skin's immune system, which may give these sleepy viruses a chance to sneak past our defences and come out to fight.[10] Although the plural of anecdote is not data, it may be no coincidence that my shingles described in Chapter 6 erupted in the lead-up to very stressful exams. Some of my colleagues would argue that it may have appeared randomly, or that my revision diet of pizza may have played a role. But with what we understand about the effects of mental stress on the skin, it shouldn't be surprising that the mind is often the gateway to physical disease.

If all of this information is now causing you mental stress, take heart that it also confuses many scientists working in the field. If anyone (or any business trying to sell you something) attempts to categorically explain the relationship between the brain and the skin, be suspicious. The brain and skin aren't on–off machines. They are complex, dynamic, fluctuating environments. All that we can confidently assert is that psychological stress definitely affects the skin. It can worsen existing skin conditions such as eczema, psoriasis, acne, alopecia and pruritus (the medical term for the symptom of itching) and it can give opportunistic microbes a foothold. It's likely, then, that at some point your skin, too, has felt the stresses and strains of life.

It is of course very important to seek medical help to treat the physical aspect of a skin condition, but a flare-up can also be a warning about our mental state, telling us that we're stressed or allowing too much emotional and mental pressure to build up. We live in a world of performance targets, airbrushed magazine models and the constant advertising of 'perfect' lives on social media. It can feel like we're playing a constant game of catch-up; the struggle to improve ourselves and our bodies both in the workplace and at home. As stress and its symptoms are very individualized, its cure can take a number of different forms. It could involve simply taking a step back from a commitment, setting aside time in the week to rest and meditate, seeking cognitive behavioural therapy (or other relaxation methods that work for an individual) or chatting to a doctor. In one study following patients receiving light therapy for psoriasis, those also undertaking cognitive therapies such as meditation needed 40 per cent less light therapy to eliminate their psoriasis.[11] Imagery, often combined with hypnosis, is also effective. I remember a patient with dry, itchy eczema saying that during flare-ups she imagined herself being splashed by a passing car on a damp, drizzly English afternoon. The wet, cooling effect of this conjured image slowly began to dampen down the itchiness of the eczema, which then sped up the healing process. Through various means, the effects of mental stress can lead to physical disease, and while creams and medications can work at putting out fires, true treatment needs to heal the root cause. Reducing long-term stress is an invaluable part of staying comfortable and happy, let alone having healthy skin. It is possible, and in some cases necessary, to treat your skin with your mind.

For many, the effects of stress on the skin are slow and subtle. But there is one near-universal human experience where the innermost workings of our mind flash up on our surface: blushing. We all know the situation. You're asking a question in a meeting but by the time you realize that the answer is blindingly obvious and has already been discussed, the words have already tumbled out. You feel embarrassed. Your face starts to feel prickly, clammy and warm. You feel as though all eyes are gravitating towards you. Someone helpfully says, 'You're going red!' You turn a darker shade of scarlet.

When we recognize that a particular situation is embarrassing, our body releases adrenaline (one of the chemicals of the fight-or-flight response). This dilates blood vessels, drawing blood to the face, ears and neck. Blushing is usually restricted to these areas, which differentiates it from 'flushing', which can affect other areas of skin (such as the torso, hands and feet) and is usually caused by medications, alcohol or an underlying disease. There are probably many other molecules and receptors involved in blushing but we surprisingly know very little about the science behind it. That's partly because it is difficult to measure. Blushing occurs in all shades of skin colour, but it is evidently more visible in light skin. A black friend remarked: 'My sisters or my mother can pick out my blushing from a mile, but it's much harder for those who don't know me to detect it. If a tree falls in a forest and there's no one there to hear it, does it make a sound?' This tellingly shows that a crucial aspect of the blush is its detection by another human. Even if you don't like the attention, the visual recognition of a blush by another person is actually important: the skin is communicating.

One of the great enigmas of blushing is why humans even have this ability in the first place. Scientists, psychologists and sociologists find blushing fascinating. In *The Expression of the Emotions in Man and Animals*, Charles Darwin observes that:

> 'Blushing is the most peculiar and most human of all expressions ... We can cause laughing by tickling the skin, weeping or frowning by a blow, trembling from the fear of pain, and so forth; but we cannot cause a blush ... by any physical means – that is by any action on the body. It is the mind which must be affected. Blushing is not only involuntary; but the wish to restrain it, by leading to self-attention actually increases the tendency.'

Darwin saw blushing as uniquely human, representing an involuntary physical reaction caused by embarrassment and self-consciousness in a social environment. If we feel awkward, embarrassed or ashamed when we are alone, we don't blush; it seems to be caused by our concern about what others are thinking of us.

Studies have confirmed that simply being told you are blushing brings it on.[12] We feel as though others can see through our skin and into our mind. However, while we sometimes want to disappear when we involuntarily go bright red, psychologists argue that blushing actually serves a positive social purpose. When we blush, it's a signal to others that we recognize that a social norm has been broken; it is an apology for a faux pas. Maybe our brief loss of face benefits the long-term cohesion of the group. Interestingly, if someone blushes after making a social mistake, they are viewed in a more favourable light than those who don't blush.[13]

If you, like many, have erythrophobia – the fear of blushing – take heart from the fact that blushing has this positive connotation. It's also not usually as noticeable as you think and is often quickly forgotten. Studies show that those who fear reddening often overestimate the costs of their blushing.[14] But if blushing does strike, there are a few simple tricks to reduce the redness. The first is to relax your face. Smiling is the best way of doing this and it's been shown not only to reduce blushing but also to ease almost any social situation. The other device is to consciously draw your own attention from the blush. Take deep breaths; focus on drawing air into your lungs and gently expelling it. This is more easily said than done but it is remarkably effective when practised. Some people discover that psychologically drawing the heat away from the face is useful. This could involve 'cooling yourself down': imagining having an ice-cold bucket of water poured over you, or focusing on drawing the heat from your face into clenched fists. It's also vital to keep hydrated. It reduces the frequency and severity of blushing, with the bonus that – as we saw in Chapter 3 – hydration is good for the health of both skin and body.

Embarrassment is not the only emotional state that causes skin to turn red. Mr Stirling, my first maths teacher, was an impatient man. I had always wondered why someone who hated the presence of children would choose his job, and with his frequent outbursts of rage I wondered more at how he managed to keep it.

Instead of teaching, he would spend the first five minutes with his back to the class, drawing up a sadistically difficult problem on the whiteboard. Without a word, he would then heave his tubby frame around to face the class. Holding the board marker in a quivering outstretched arm, in a strange game of 'pin the tail on the donkey' his searching glare would settle on a random member of the class. This time he pointed directly at me.

'Boy, solve it!'

Staring up at an indecipherable mess of symbols and numbers, I had no hope. I stalled and mumbled for a few moments, though it felt like months. I could feel the warm tingle of redness rise up across my neck as, sensing twenty pairs of eyes staring at me, I began to blush. But that was nothing compared to Mr Stirling. As he shook with anger and impatience, his bald head started to glisten with sweat and the veins across his temples seemed to swell. Then the floodgates opened. His face dramatically flushed with vivid crimson, like a boil ready to burst.

'If you don't answer, I'm sending you out!'

Some people go 'red in the face' with anger when their carotid arteries, which supply the head and neck, dilate and rapidly increase blood flow to the face. This could be acting as a safety valve in response to dangerously high blood-pressure levels in the acute state of anger. Another reason could be that in the fight-or-flight response, where you would logically think blood flow would be diverted to muscles, a bright red face – the colour of danger, in nature – acts as a warning sign. Skin shouts: 'Stay away!' It was certainly a good idea to avoid Mr Stirling.

The mind also appears uninvited on the skin in the form of sweat. We experience the 'cold sweat', as it's often known, in stressful, uncomfortable, embarrassing situations and, as with blushing, thinking about sweating can make it worse. The sweat glands in skin are activated by the arousal of the sympathetic nervous system, which is activated by the fight-or-flight response.

Sometimes, though, excess sweating (hyperhidrosis) isn't psychological at all. There are some quick fixes to break the vicious cycle of sweating and the worry that accompanies – and worsens – it. This includes wearing loose clothing with sweat-disguising colours, namely white or black, avoiding triggers such as caffeinated drinks, and applying daily antiperspirant, not simply a deodorant. If regular antiperspirants don't work, it is worth applying one with higher levels of aluminium compounds, which plug up the sweat glands. These usually come as roll-ons that are applied at night, and while a common side effect is skin irritation, it's often a small price to pay.

In the early 2000s, rumours circulated that the preservatives used in antiperspirants increase the risk of breast cancer.[15] This is a myth that can be traced back to spam emails. On the balance of evidence, there is no causal link between antiperspirants (aluminium-based or otherwise) and breast cancer.[16] Short-term, toxic effects of aluminium have been harder to measure, as it's difficult to assess how much is able to get through the skin barrier. The scientific community generally agrees, however, that they are safe to use at the recommended doses. The benefit of a strong antiperspirant or sweat-absorbent pads is that, even if they bring about only relatively small reductions in sweating, this makes the sufferer think about it less, which further improves the situation.

As with stress-induced skin inflammation, if blushing and sweating are causing problems, there's no shame in seeking medical help. Psychological therapies and relaxation techniques have been shown to help decrease the anxiety that often lies at the root of these uncomfortable experiences. In extreme cases, where all other treatments have failed, there are specialized surgical procedures that can prove effective in both of these conditions, but the vast majority of cases can be resolved with much more conservative interventions. If problems are bottled up, hidden behind the barrier of one's skin, and the skin doesn't seem to do what the brain is telling it, one of the simplest and most effective strategies is talking, whether to a medical professional or to a friend.

Blushing and sweating are, in part, a manifestation of unspoken thoughts on the surface of our skin, so perhaps it isn't surprising that humans have tried to exploit this. It has long been known that skin has continuously varying electrical activity, as demonstrated by the use of its conductive powers to enable humans to interact with a smartphone touchscreen. In 1878 the Swiss scientists Hermann and Luchsinger found that the variation of electrical activity was strongest in the palms of our hands, leading them to discover that sweat, which contains water and electrolytes, was the greatest factor in increasing the electrical signals.[17] It didn't take long for scientists to realize that tiny, unnoticeable changes in the 'electrodermal activity' of skin could be directly related to a subconscious emotional state of arousal. On seeing the effects of our deepest thoughts seeping through our skin, Carl Jung, the legendary Swiss psychoanalyst, is said to have exclaimed: 'Aha! A looking glass into the unconscious!'[18]

The discovery that sweat spills our secrets led swiftly to the development of a controversial device that would affect the lives of thousands across the world: the polygraph, or lie detector. In the 1930s Leonarde Keeler (named after the polymath Leonardo da Vinci) incorporated electrodermal activity into emerging machines that measured blood pressure and heart rate, to try to detect deceit.[19] In 1935, results from Keeler's polygraph were used for the first time as evidence in a US court and after a Wisconsin jury was swayed by polygraph results, Keeler announced 'the findings of the lie detector are as acceptable in court as fingerprint testimony'.[20] This would have been a triumph for justice and science if the polygraph was 100 per cent accurate, but it is not. It measures arousal, but cannot differentiate between emotions, whether they be guilt or anger. Humidity, temperature and medications are among many factors that can alter electrodermal activity and skew the result. It's also possible to cheat lie detectors, while some people with sociopathic-like personality disorders (commonly referred to as 'psychopaths') don't experience any emotional arousal during interrogation. Although lie detectors are now not permitted as evidence in court in the USA and most European countries, the historical replacement of the jury with this controversial machine has had dreadful consequences. In 2006, Jeffrey Deskovic was exonerated after sixteen years in prison when DNA evidence finally proved that he had been wrongly convicted of the rape and murder of a fifteen-year-old girl. His conviction had been almost entirely based on a failed polygraph test followed by a false confession.[21]

On a more positive note, modern research is finding that measuring electrodermal activity could actually combat stress. The Pip, a small handheld device that monitors electrodermal activity eight times a second, relays the information on to the screen of a smartphone or computer and, with relative accuracy, lets you know your current state of stress. Combined with positive reinforcement in the form of games with points scored when you reduce electrodermal activity, these devices can be effective at reducing stress levels. Such 'biofeedback' therapy calms down minds and soothes bodies, with the potential to improve the symptoms and progression of various physical maladies, from heart disease to migraines.

Electrodermal activity has also been used to study another remarkable and mysterious skin phenomenon: frisson. When listening to the climax of a rousing piece of classical music, or a pop song that evokes particularly special memories, it's possible that you feel a warm, pleasurable wave of chills run up the skin of your spine, sending a shiver of goosebumps over your neck, face and arms. If so, you are in the two thirds of people who experience frisson, or 'aesthetic shiver'. This overwhelming feeling of aesthetic tingles, when the mind has complete control over the skin, can be triggered by a moving scene in a film or a particularly beautiful painting, but is most easily brought about by music.[22] I had always assumed that 'emotional' people would experience more musical chills, but research suggests that the main predisposing factor for frisson is cognitive engagement with the music. If a composer wants to excite the skin of the listener, they need to be playful. We now know, thanks to musical scientists, that frisson is brought about when our expectations are violated, and then resolved, in a positive way. Psyche Loui, a neuroscience researcher at Wesleyan University, also a violinist and pianist, was intrigued by these feelings. Reviewing the evidence, she found that changes in melody and pitch, as well as a slight dissonance that is quickly resolved, toy with our expectations.[23] As we develop, our brain constructs rules about how songs are composed, particularly in relation to cultural musical norms, so if a piece sticks too close to the formula it is dull but if it strays too far it becomes dissonant noise. When we hear playful melodic tension, our brain is teased, and our skin feels it.[24]

I first realized that these shivers were part of a game of anticipation and violated expectations when I was watching television with my family in 2009. It was the auditions of *Britain's Got Talent*, and the TV audience was introduced to Susan Boyle. After less than a minute of interview we found out that this forty-six-year-old Scottish spinster was unemployed, living on her own with her cat Pebbles and had never been kissed. As she entered the stage, Susan was greeted by howls of mockery and wolf whistles from the live audience. She was presented as the ugly opposite of everything a female singer was supposed to be. She was not meant to be good. When the music started and she began slowly, and wonderfully, to sing 'I Dreamed a Dream' from *Les Misérables*, the audience was stunned into silence – then burst into rapturous applause. Not one member of my family was left without the shivers.

Frisson is not only about surprise, however. We get conditioned to feel it, with people reporting that they always get goosebumps at particular points of favourite, meaningful songs. The pleasure felt on the skin during this ecstatic listening experience starts in the brain: the musical trigger unlocks the release of opioids and dopamine (a key molecule in the brain's reward pathway) into the same pathway activated by sex, food and recreational drugs. (When people are given naloxone, an opioid blocker used to reverse the effects of a heroin overdose, they can't feel frisson.) These chemicals make this skin sensation addictive and their happiness-inducing qualities partially explain why listening to good music with friends consolidates relationships and bolsters empathy and altruism.

The mind affects the skin, but the skin also directly affects the mind. Skin is a book, the only part of us that is exposed to the outside world and, for better or for worse, it forms part of the first impression we give to others. We can feel both defined and confined by our own skin. How we think others perceive our skin affects our minds in the short and long term. The existence of the multi-billion-dollar cosmetics industry is visible proof of how crucial our skin is to identity, but it is often most acutely felt by those with visible skin diseases. In the American novelist John Updike's memoir *Self-Consciousness*, he gave a whole chapter to his physical, mental and social battle with psoriasis. There's a reason why dermatology is one of the few medical specialities that has needed to develop a 'Life Quality Index' for patients.[25] This questionnaire works out the emotional, social, sexual and physical burden that often accompanies a disease of our outer organ.

A startling, but often overlooked, example of a condition of skin affecting the mind is the recent finding that one in five acne sufferers in America and Britain has considered suicide.[26] Acne vulgaris is very common and usually erupts with the hormonal changes of the transition from childhood to adulthood. It comes at a formative time in the development of friendships, romantic relationships and first impressions when starting college or a career. With or without bullying, if it is not taken seriously acne can have crippling effects on confidence, social development and mental wellbeing.[27] Acne is also exacerbated by stress, dragging sufferers into a vicious whirlpool of episodes of stress that cause flare-ups, which further pile on the depression and anxiety. A study at Stanford University found that college students were much more likely to have breakouts of acne in the build-up to examinations.[28] During emotional and mental stress, levels of cortisol and testosterone increase, which stimulates sebum production in the skin and oils the acceleration of acne. All of this is compounded by the fact that the understandable temptations to pop or scratch the wretched spots can cause permanent scarring, dragging the vortex of despair into the rest of the sufferer's life. I remember seeing a twenty-six-year-old patient who, after having had clear skin for almost ten years, erupted in papules and pustules in the weeks leading up to her wedding. The resulting feelings of shame eventually forced her to suspend the wedding until her acne was treated. In many ways, acne is more of a psychological disease than a physical one. The pimples of acne, wrongly associated with poor hygiene, bring on bullying in formative years that can stunt social and psychological development. Even for those acne sufferers who come out of their teens without physical scars, the emotional and psychological scars from this formative period of social development can last a lifetime. Acne is too often dismissed as a common, minor condition and the spots are trivialized. It needs to be taken much more seriously in society and within the medical profession, where we frequently see how people's lives have been changed by the condition.

One muggy summer's afternoon I sat in on a dermatology clinic in an ethnically diverse area of Birmingham taking a medical history from an elderly Irish woman. She spoke of attempts at ending her own life due to rosacea giving her a 'red, knobbly, hideous face'. She used to be a model. The very next patient was a young woman of Pakistani descent who had vitiligo. Her depigmenting skin condition had resulted in asymmetrical patches of white skin on the left-hand side of her face. She had severe clinical depression and believed she would never marry due to her appearance. Searching through medical literature, I found studies showing that in both conditions almost half of sufferers have reported depression.[29,30] One of my friends, a surgeon, constantly dismisses dermatology, claiming that it doesn't deal with life-threatening conditions. But I would argue that most skin diseases, particularly if they are visible, can be life-ruining.

The most dramatic manifestations of mind on skin can be found in psychiatric conditions. My first introduction to the psychiatry–skin connection was during a conversation with a retiring dermatologist in Oxford, who told me a story about a patient she met soon after she started her job as a young doctor. Jack was a gaunt, gangly young man wearing baggy grey overalls speckled with paint, and the first patient on the day's list.

'Take a seat,' she told him as he entered the consulting room. 'How can I help you?'

'Well, you see, I've got bugs. I started to feel an itch – no, a crawling under my skin. I used to do gardening, you see, and maybe some kind of bug is under there and they're multiplying. All down my arms, and here . . .' Jack pointed to various areas across his chest and stomach. 'I'm done in. I can't sleep, I can't concentrate at work. I was working in a fancy garden, you see. Some weird, foreign bug has got into my skin and is laying eggs – you can see them right now! Little black insects crawling under the skin.'

The dermatologist carefully looked over at the area of smooth, unblemished skin to which Jack was pointing and failed to see anything. Before the doctor could ask any further questions, Jack fumbled in the loose pockets of his overalls, then brandished a small glass jam jar. It seemed full to the brim with cheese. He plonked it on her desk as though producing a royal flush.

'This proves it, doctor! This is what I showed the first doctor but he wouldn't listen to me!'

On closer inspection, the container was packed full with tiny, cheesy, green-tinged brown flakes.

'It's my skin! Take it to a lab – they'll prove that I'm infested. I want people to stop ignoring me!'

The baffled junior doctor inspected Jack's skin, finding nothing aside from scratch marks caused by his itching. She reassured him and took the samples to the laboratory. When the results came back, there was absolutely no evidence of an infestation. The skin flakes were old and a little smelly, but otherwise normal. She was soon to learn that Jack had a psychiatric condition called 'delusional parasitosis'. This is when a patient is convinced, even in the face of overwhelming evidence to the contrary, that their skin is infested with insects. They have a crawling feeling under the skin, known as 'formication'. Many patients are so certain that they bring containers of their own skin flakes to prove the existence of the bugs, known as the 'matchbox sign'.

In Jack's case, the delusional parasitosis was a stand-alone disease with a purely psychiatric cause, but it can also be found in patients with illnesses such as diabetes and cancer and can also result from medications and recreational drugs, most notably cocaine. A more correct term for the disease is 'delusional infestation', as these days insects are being replaced by the objects of our technological world: increasingly, cases feature individuals believing that nanotubes, microfibres and even tracking devices lie underneath their skin.

One method used in the study of delusional infestations is the remarkable rubber-hand illusion, introduced in Chapter 6.[31] With both hands resting on a table, the participant's left hand is hidden behind a screen, with a realistic rubber hand placed near their hidden left hand, but in full view. The researcher then starts stroking the index finger of the rubber hand whilst stroking the index finger of the participant's hidden left hand at the same time. After about a minute, the brains of roughly two-thirds of people trick them into thinking that the fake hand is their own. This creates an awkward battle between 'bottom-up' perceptions from our senses of sight and touch and 'top-down' knowledge that the rubber hand doesn't belong to us. Patients with a delusional infestation are highly responsive to this test, easily believing that the rubber hand they can see being stroked is their own. This suggests that there could be an error in the ability of these people to identify and interpret 'reality' from a combination of different bottom-up sensory inputs. Equally, top-down functions of cognition may be altered, as shown by itching that becomes markedly worse in these patients when insects are brought up in conversation. From what initially appears to be a disease of the skin, we are given unprecedented access into the peculiar workings of the human brain. As delusional infestation is a psychiatric condition it needs to be treated delicately by a psychiatrist or a psychodermatologist.

While we wonder at rare and remarkable psychiatric conditions, it is actually the ones that seem harmless that can be the most

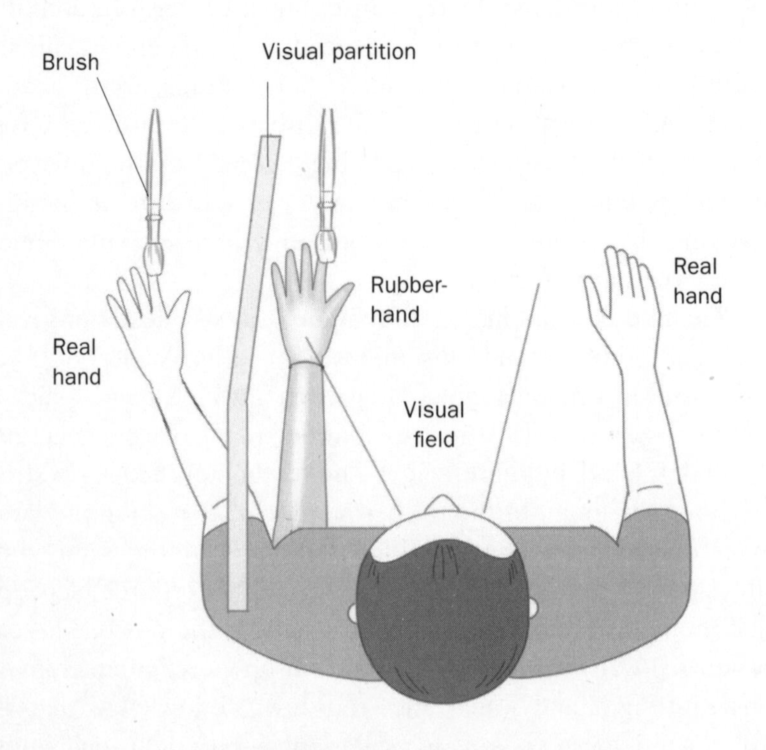

RUBBER-HAND ILLUSION

devastating. It's easy to trivialize obsessive–compulsive disorder (OCD): 'Michael has terrible OCD! He can't start working until he's made sure that all the office fire escapes aren't blocked!' But for those diagnosed at the severe or persistent end of the OCD spectrum, the disease is a window into one of the most delicate, complicated and dark corners of the human condition.

Obsessions are intrusive thoughts that persist despite attempts to prevent them, while compulsions are rituals that a person feels that they have to perform. Compulsions often manifest themselves in destructive ways, more often than not on the skin. In fact you are roughly ten times more likely to find someone with clinical OCD in a dermatology clinic than on the street.[32] OCD is often seemingly never-ending: the obsessive thought (such as having dirty, contaminated hands) causes distress, which is temporarily relieved by the compulsive act (hand-washing), with the process sometimes repeating itself hundreds of times a day.

A number of other skin manifestations of OCD are labelled, using Ancient Greek, as either 'tillomania', for picking or pulling, or 'phagia', for biting. Compulsive hair-pulling is known as trichotillomania and nail-biting as onychophagia. Dermatophagia, eating or chewing the skin, is much less common than nail-biting but its prevalence is dramatically increased in those with OCD, impulse-control disorders and autism. The areas most commonly chewed are around the nails, along the knuckles and on the inside of lips. The damage done by this psychological condition can be socially isolating and physically devastating. Skin-biting damages the outer barrier and leaves the body open to infection; compulsive nail- or hair-eating can pose more of a risk to the gastrointestinal tract. An extreme consequence of compulsive hair-eating is known as 'Rapunzel syndrome', where the tail of a hairball in the stomach reaches down into the intestines, causing a potentially fatal bowel obstruction. In 2017, a 16-year-old British girl died from Rapunzel syndrome when an infected hairball perforated her stomach.

Trichotillomania is a condition often completely bypassed by medical students and doctors, but its first description, by Hippocrates, led him to advise physicians to routinely look for hair-pulling in their patients. He had come across Thasos, a woman in the throes of inconsolable grief, who 'groped about, scratching and picking out her hair'. Contrary to what you might think, 'pulling your hair out' is not usually a response to extreme emotional stress; most patients displaying this behaviour slowly pull out their hairs during routine daily activities.

Body dysmorphic disorder (BDD) is also classified on the OCD spectrum.[33] What seems to be a tiny pimple on the skin to others may seem like Mount Vesuvius when the sufferer looks in the mirror. As opposed to vanity, where the concern is to look more beautiful, those with BDD obsess on reaching an apparent norm. While the condition can involve any aspect of a person's appearance, the skin is a crucial issue in 73 per cent of patients with BDD. In contrast to other conditions on the OCD spectrum, BDD also has higher rates of depression, social avoidance and suicide. There is a vast array of ways of managing conditions on the OCD spectrum, including distraction techniques and therapies designed to slowly expose the patient to the trigger, gradually desensitizing them. However, obsessions and compulsions are by their nature very hard to stop, and those with severe OCD are some of the hardest people to treat in the whole field of psychiatry.

Whether at daily consultations at the family doctor's or at the clinics of super-specialized psychiatrists, the skin is often a battleground of one of the most complex and impenetrable frontiers of medicine: the unexplained physical symptom. I once saw a man who complained of tingling and numbness across the skin of both of his legs, accompanied by itching all over the skin of his upper body. The bilateral leg symptoms raised alarm bells concerning potential spinal damage or disease. After a series of scans that came back negative and a further consultation, his story began to unravel. He was working three poorly paid jobs, caring for a wife with terminal cancer and trying to raise two children. The pressure had wrought upon him overwhelming anxiety, and the psychological distress had converted into physical symptoms on his skin. This phenomenon, where struggles of the mind manifest in the body, is called somatization. This may have been heightened by the extreme stigma of mental illness in his culture and the pressure for him to 'be the man'. These physical symptoms on his skin were completely relieved by psychological treatment: a course of cognitive behavioural therapy that managed to get to the root of his stress and psychological issues, and ultimately dealt with his physical symptoms. Skin is a physical stage on which the enigmatic actors of our cognitions, behaviour, mood and perceptions are always performing.

In 2013 a team of Californian dermatologists published an unusual case report.[34] Janice, a fifty-one-year-old woman, came to accident and emergency with sudden muscle weakness on her right-hand side, loss of memory and an inability to formulate words properly. If you had to put your money on a diagnosis at this stage it would have been a stroke. Whilst Janice was being assessed, the team noted her acne, with a few healed scars on her face. That wouldn't necessarily be the first thing an emergency doctor would think about, but their gaze was drawn to a small gauze dressing that covered a scab on the hairline above Janice's forehead. When one member of the medical team tentatively pulled back the gauze, imagine their surprise when they could literally peer into the diagnosis. It turned out that Janice had been repeatedly picking at the skin on her forehead with a sewing needle. She had previously developed a painful 4x2cm ulcer and, although she knew that she was harming herself, the compulsion to pick was irresistible. Janice picked at it over a number of months, slowly drilling down through skin, connective tissue and muscle, then eventually creating a small hole in the skull. The resulting damage to the brain was the cause of her neurological symptoms.

This case is the most literal example of travelling from the skin to the brain. Janice used the piece of gauze to hide a physical tunnel linking these two organs. For a great many other people these tunnels are invisible but feel just as real, and can propagate cycles of depression, social isolation and stress. Just as the mind and skin are intertwined, so our mental and physical wellbeing go hand-in-hand, and sometimes we can't have one without the other.

8

Social Skin

The meaning in our markings

'Taia o moko, hei hoa matenga mou'
(Inscribe yourself, so you have a friend in death)

(MAORI SAYING)

FROM THE BEACH in the tiny port of Russell, near the tip of New Zealand's North Island, it is possible to make out the headland, where in 1840 the Treaty of Waitangi – which declared British sovereignty over New Zealand – was signed by Maori chiefs and British delegates.[1] But the two alien civilizations were starting to rub against each other. The sleepy village of Russell, nestled in the quiet Kororāreka Bay (translated pleasantly as 'the sweetness of the penguin') belies its violent history. As the first European settlement in New Zealand, Russell's reputation for pirates, smugglers and prostitutes earned it the moniker 'hellhole of the Pacific', and in the mid-1840s Russell became the frontline of the First Maori War, between the British settlers and the indigenous population. A flagstaff, which once flew the British flag, still looms over the settlement from a nearby hill. It is, in fact, the fifth flagstaff to stand here, as Maori warriors repeatedly felled them in the years following the treaty, when the village was continually ransacked.

Around two hundred years ago, the five-year-old son of one of my ancestors was playing in the shallows of Russell's beach with a local Maori boy. In childhood obliviousness they had struck up the best of friendships, despite being at the epicentre of a raging cultural war. While the boys were splashing around and throwing sand at each other, the water became choppy and the Maori boy was suddenly dragged out to sea by a current. My juvenile relative frantically waded out to try to save him. Neither could swim, and both boys drowned. They were buried together in the town's Anglican Christ Church, New Zealand's oldest. When I visited Russell I found out that the local school had named a swimming prize after the boys, which felt particularly moving as I had become an obsessive open-water swimmer.

We all love to find out about our family history, as we feel it may tell us something about ourselves; just look at the exponentially growing interest in online ancestry sites, the many celebrities uncovering their past on the BBC's *Who Do You Think You Are?* TV series and the increasingly easy access to DNA testing. My family is of European ancestry so our story has for centuries largely been written on paper (albeit often with questionable grammar and inconsistent name-spelling). The family history of the Maori, however, is written on their skin.

Though getting inked up at your local tattoo studio is certainly not pain-free, spare a thought for those who have undergone *Tā moko*, the traditional art of Maori tattooing. In the past, instead of inserting ink using a needle, the skin of a Maori would be cut open with an *uhi* (a chisel made of albatross bone), then pigments from fungi and ash would be placed in the wound, which was left open to heal slowly. The recipient's face was often so swollen that they had to be fed through a funnel for several days. Over the course of their lives, a man would gradually have his whole face covered, whereas a woman usually had a signature tattoo on her lips and around her chin. In 1769, when the *Endeavour* and its tattooless European crew first made contact with the locals, Captain James Cook quickly recognized that the intricate lines of *Tā moko* combined beauty, meaning and individuality.

> The marks in general are spirals drawn with great nicety and even elegance. One side corresponds with the other. The marks on the body resemble foliage in old chased ornaments, convolutions of filigree work, but in these they have such a luxury of forms that of a hundred which at first appeared exactly the same no two were formed alike on close examination.[2]

With no books or papers, most Maori had their story written on their skin. In Rotorua I was able to speak to a Maori chief about the meaning of each line. He had been closely involved in the *Tā moko* revival in New Zealand over the previous few decades. 'If you know the language, you can read me like a book,' he told me, grinning. His smile animated the intricate whorls on his skin, which hugged the contours of his lips and cheeks. 'Generally speaking, if you're high up enough to have *Tā moko* on your face, your rank is marked on your forehead and around your eyes, your birth status is drawn on to your jaw, and the lands and wealth you have acquired go on your chin. You English would hate it, knowing if someone's from old or new money! On top of your nose roughly shows your educational background. Then you also have a unique signature.' He pointed to a pattern between his upper lip and nose. 'An incomplete pattern is a sign of shame; you haven't had the guts to last the cutting.'

The Maori tattoo is akin to having your genealogy, CV and bank balance written on your face. Like the best stories, Maori tattoos combine beauty with bravery, function with form. *Tā moko* flatters the face and cheek bones, and draws attention to the eyes and lips. It is meant to be intimidating, yet attractive. As each facial tattoo is as personal as a fingerprint, in some of the many treaties with the British a Maori chief would have his *Tā moko* drawn on to the document in lieu of a signature.

Given that *Tā moko* represents a person's life and family story, it is not surprising that the tattoos were deeply sacred. When a Maori warrior died, his tattooed head (*mokomokai*) would be smoked, then dried in the sun to preserve its patterns. Maori archives thus comprised the skin of their ancestors. Even in war, it was customary for the winning tribe to hand back the heads of the enemy to their respective families, and *mokomokai* were often exchanged during peace treaties. The arrival of the British in the 1800s not only led to the death of *Tā moko*, due to the Christian disapproval of tattoos, but meant *mokomokai* became very scarce, with European collectors quickly developing an insatiable appetite for these heads. During the 1820s, demand was so high that some Maori killed each other to profit from the trade.

The legacy of this gruesome history continues to the present day, and has had implications for two of the universities I have studied at. The University of Birmingham Medical School, one of the oldest and largest in England, was bequeathed thousands of historical artefacts and anatomical curiosities from all over the British Empire by wealthy alumni and donors in the eighteenth and nineteenth centuries. In 2013 a delegation from the Museum of New Zealand Te Papa Tongarewa in Wellington was dispatched to Birmingham to retrieve a number of tattooed Maori heads which, still being sacred to the Maori, were to be repatriated. Two of my lecturers, Professor Jonathan Reinarz and Dr June Jones, helped organize a ceremony at the university marking the beginning of their return, followed by a funeral in New Zealand. A similar ceremony was carried out in 2017 at the University of Oxford's Pitt Rivers Museum, whose treasure trove of tribal trinkets from the days of Empire contained *mokomokai*.[3] Today the sacredness of the Maori tattoo also applies to living bearers, with many New Zealanders outraged at non-Maoris in the public spotlight – such as the singer Rihanna and the boxer Mike Tyson – appropriating the sacred markings on their own skin.

Wherever you go in the world and no matter how far you go back in recorded history, humans have always tattooed themselves. Indeed, humans are unique in permanently marking their bodies to communicate with others. To make these permanent marks, we

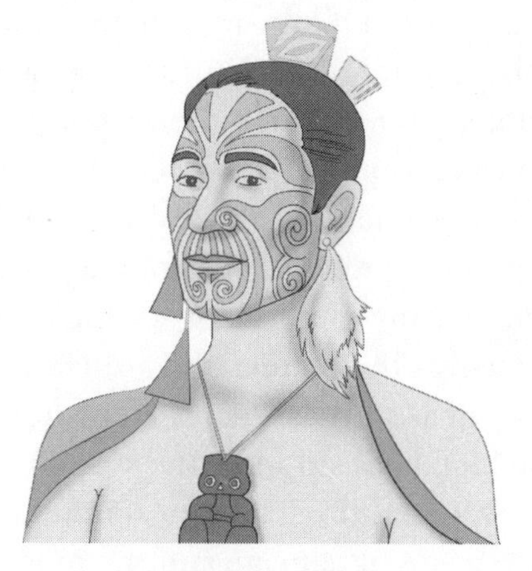

MAORI TATTOO

exploit beautiful, but little-known, intricacies of our skin, where the physical and social are sometimes inseparable.

Imagine that as you turn this page your index finger scrapes along the fine edge of the paper, which slices deep into your skin. Blood starts to ooze from the wound and it's surprisingly painful although, as it's just a paper cut, the pain is something you would probably find hard to admit. Have you ever thought about how your body responds to this attack? Your skin immediately springs into action and begins to compose a four-movement symphony. The body's first priority is to stop the bleeding, a process called haemostasis. When the paper edge carves through the tiny blood vessels in your dermis, the local pain receptors in the skin cause these vessels to spasm and constrict, reducing blood flow to your new cutaneous crevasse. Within a couple of minutes the emergency services – platelets – swing into action. These disc-shaped cells, much smaller than red or white blood cells, usually drift unobtrusively along our bloodstream. However, when they arrive at a wound site they stick to the collagen in the dermis, as well as to the damaged inner layer of the vessel, and become activated. On activation, the platelets rapidly morph into awkward, irregular shapes to help them interlink and stick to one another as strongly as possible, forming a lump. They then release a cocktail of molecules that both causes further constriction of local blood vessels and attracts more platelets to the ever-expanding lump. This platelet plug begins the process of coagulation, where a number of proteins – called clotting factors – work together in an intricate chain reaction to cover the platelets in a thick mesh called fibrin. This 'haemostasis phase' all happens in a matter of minutes.

Now that the blood has stopped, the second stage, the 'inflammatory phase', can kick into gear. Our immune army – both the local garrison at the site of damage and the more specialized cells from the rest of the body – is called in for two roles: the military imperative of killing bacteria that enter the body through this breach in your skin's defences and the disaster-relief, clear-up task of removing debris and destroying dead cells. Over the next few days the inflammatory phase transitions into the 'proliferative phase', where our skin's builders, the fibroblasts, get to work. These start to rebuild the wreckage by producing new collagen and proteins that aid the healing process.

In wounds that are wider and more extensive than a paper cut, the skin recruits a remarkable team of especially brawny builders, the myofibroblasts. These travel to the wound edges and contract, pulling the wound closer together at a speed of almost 1mm a day. If needed, molecules released around the wound site are able to promote normal fibroblasts up the ranks, transforming them into myofibroblasts to join the effort. During this period, new blood vessels also start to grow into this area of new connective tissue, filling the wound site. This miscellany of new cells and blood vessels, known as granulation tissue, is disorganized and messy but it forms vital scaffolding for the rebuilding of our epidermis.

Keratinocytes from the *stratum basale* – the basal layer of skin stem cells that continuously repopulates our outer barrier – now slowly crawl from the wound edges across this bed of new tissue.

The 'maturation phase' completes the symphony of wound healing, with the chaotic granulation tissue continually rearranged to align with the normal tension lines of our skin. Any cells or blood vessels no longer needed are destroyed over the coming days and weeks by programmed cell death. The beautifully complex, and often overlooked, process of wound-healing regenerates all the layers of skin in your paper cut so that soon enough, when you look down at your index finger, you'd never know it happened. However, wounds where the edges are further apart than your paper cut will usually leave a visible scar – a block of collagen that, while not carrying out all the diverse functions of skin, at least forms a permanent barrier.

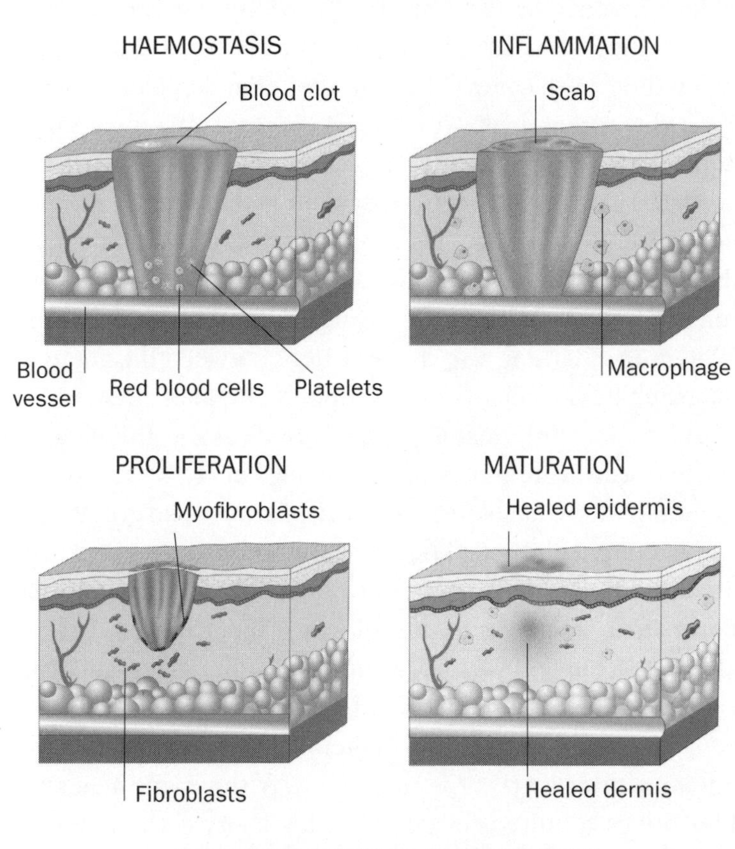

THE FOUR-MOVEMENT SYMPHONY

Wound-healing often leaves permanent marks on the skin. Perhaps the most rudimentary example of skin as a social organ is that, as humans, we have intentionally turned this damage into dialogue.

A young boy is running across the backs of a herd of bulls. If he succeeds in getting to the other side of the herd without falling to the ground, he becomes a man. This coming-of-age ritual is unique to the Hamar tribe in southern Ethiopia. The ancient customs of the Hamar, one of the Omo River tribes, have been little exposed to modernity until recent years. While the boy is running across the backs of the somewhat surprised cattle, his sisters take part in a ceremonial battle. They begin to taunt the men of the village, saying that they will never let their brother go. In response, the men whip them with canes. The young women don't let out a single sound as the ceremonial whipping leaves open wounds on their backs, blood dripping on to the African dust. The substantial scarification that is left on a woman's back from this seemingly horrific process subsequently becomes a proud symbol of belonging, one she shows off for the rest of her life. The criss-cross scars tell a story of strength, bravery and intense loyalty to her family and community. This crude marking also creates a debt between the young man and his sister; he now has a duty to support and look after her.

In the remote interior of Papua New Guinea, the Kaningara people take scarification to another level. Every five years or so the young men of the village undergo a rite of passage so gruelling that survival is not guaranteed. They first spend two months in a spirit house, with their family gathered outside whispering stories of their ancestors, while inside the elders subject the boys to ritual humiliation. When they emerge into the light, dazed and emaciated, the crocodile cutting begins. The Kaningara, who live on the banks of a crocodile-infested river, believe that they descend from a divine apparition of this riverine beast. Without any anaesthetic, an elder drives hundreds of deep cut lines across the chest, back and buttocks of the initiate with a stick of sharpened bamboo.

River mud is then forced into the bleeding wounds, which slows down the healing process. The result is a body covered in hard, raised keloid scars that stand like a ridge on the skin. These pronounced scars are caused by the prolonged healing process stimulating fibroblasts in the dermis to produce excessive amounts of collagen that ultimately creates an overgrowth of scar tissue. Those who do not die from shock or infection are proud of their rough, bumpy, crocodile skin as they believe they are now endowed with the strength and ancestral blessings of this fearsome reptile.

Permanent body markings, whether drawn by scars or dye, paint a picture on the body of the subject not only in terms of the design of the markings themselves, but also of the method used to make them. Given the excruciating pain associated with the receipt of tribal tattoos, it is not surprising that they play a sacred role in rites of passage. Coping with the pain demonstrates that a young man is now ready for the tribulations of battle or that a woman is strong enough for childbirth. Markings engraved on human skin tell the story of what the initiates have been through to have them, as a kind of foretaste or prediction of what they are now capable of achieving as warriors, or adults, or mothers.

The gradual adoption of permanent skin-marking in the West, leading to the phenomenon where roughly a third of Americans and Brits aged twenty-six to forty sport at least one tattoo, takes us back to the late nineteenth century. The first recorded tattooist in Britain operated in the port city of Liverpool in the 1870s. Tattoos were initially considered novel and exotic, and came at extreme expense, being first a hit with the upper classes (and royalty, including George V and Tsar Nicholas II of Russia) until an inventive American made the practice available at low cost. In 1891, in a small shop in Chatham Square, New York, Samuel O'Reilly patented the world's first tattoo machine.[4] It was almost completely based on Thomas Edison's design for a rotary electric pen; what Edison had envisaged for a pen that duplicated handwritten documents became the tool for the mass-engraving of human skin. It was only appropriate that Edison went under the needle himself, receiving a quincunx, a pattern of five dots as seen on the side of dice. The tattoo machine – based, as it is today, on a rotary device or an electromagnet – industrialized the world's oldest form of written communication.

When I'm rushing around the ward on particularly hectic days I use a ballpoint pen to scribble a note on to the back of my hand as a reminder to do something; the skin becomes a busy person's Post-it note. What is already an inefficient means of capturing thoughts and ideas is completely impractical for a doctor, however, as after a couple of hand washes all traces of the ink have disappeared. So why does a note in ink on my hand disappear in a few hours, when tattoos can still be detected on a 5,000-year-old mummy? We shed a million skin cells into the air each day, so why do tattoos never leave us? The unexpected answer lies in the skin's remarkable immune system.

Imagine for a moment that you are in a tattoo artist's chair and she is about to start inking the 'N' for 'No Regrets' on to your left shoulder. A needle impregnated with black ink penetrates your skin, pushing beyond the outer epidermis, the layer where my ballpoint pen had made its temporary mark, into the deeper dermis of the skin. The needle fires into your skin at roughly a hundred times a second, intentionally causing many tiny wounds, alerting the body to damage. Rather than being injected, the ink is sucked up by capillary action in the dermis, the ink particles waiting for your immune cells to rush in to the damaged area. In the same way that they engulf bacteria, macrophage ('big eaters', from the Ancient Greek) cells detect the pigment particles as foreign and try to eat them up. But their eyes are bigger than their stomachs, and many of these macrophages end up stuck, with the pigment trapped inside them. While the top layer of skin is always renewing itself, the inked-up cells in the dermis stay there for the remainder of our lives, locked in for ever like intricate fossils on a cave wall. We essentially create an infinite infection. So if you sport a tattoo, spare a thought for the little fellows who went into battle thinking they were fighting an infection but were instead fated to spend the rest of their days embedded in your skin-based art.

In 2017 a thirty-year-old Australian woman went to the doctor with lumps growing in both her armpits.[5] Further scans revealed more of the unusual lumps in her chest. It looked like a lymphoma – a form of blood cell cancer – but the doctors were surprised that she had none of the other symptoms of fever, night sweats and weight loss. A biopsy of one of the lumps did not find cancer, but instead extracted ink. The lumps were actually inflamed lymph nodes, enlarged by a battle between immune cells and tiny pigment particles from a tattoo on her back that had been drawn fifteen years earlier. As our surface is intricately linked to the rest of our body, it's apparent that tattoo ink also travels along these subterranean paths, colouring the inside as well as the outside. This may not necessarily mean bad news, however, and some scientists are trying to exploit the immune system's penchant for eating up these inks. A 'proof of concept' study in a laboratory at Rice University, Houston, in 2016 demonstrated that nanoparticles tattooed into the skin can be taken up by immune cells, which are then deactivated.[6] This could pave a path for treating the immune system's self-reactivity in autoimmune diseases such as multiple sclerosis.

Given that tattoos break the skin barrier and introduce metallic salts and organic dyes into the body, it is not surprising that adverse reactions are found in a small proportion (around 10 per cent) of people who have this body art, mainly from infection and allergic reactions to the pigment.[7] I once tried to refer a patient for an MRI scan, but he refused. He then told me that a previous scan had caused a black winged tattoo on his chest to blister and burn. It is rare, but not unheard of, for the magnet in an MRI to attract the metal particles in the pigment of large black tattoos, particularly iron oxide, the effects ranging from tingling to second-degree burns.[8]

But are there any long-term health effects of permanently

placing pigments in our body? In a 2017 French and German study, special X-ray technology detected that tattooing causes minutely small metal 'nanoparticles', including titanium dioxide, to be deposited in the skin.[9] Some of these are classified as carcinogenic and can also have toxic effects on some organs, such as the liver, though at the moment there isn't much evidence to suggest that tattoos lead to cancer.[10] If you are worried about the effects of tattoos on your health, the sensible answer is not to avoid them completely but just to think before you ink.[11] Recognize that despite the popularity of tattoos, there are risks involved with injecting metallic dyes – some of which we may not yet understand fully – into our skin.

Now imagine that, the artwork complete, you examine your new masterpiece in the mirror. To your horror, you realize that the tattooist is no wordsmith and has written 'No Regerts'. Regret suddenly takes on a new meaning for you. Whether or not the tattooist makes a mistake, about one in seven people regret undergoing permanent skin-marking, and many take active steps to remove it.[12] While watching an England football match, an Indian school friend of mine noticed a glaring error on David Beckham: a large tattoo along his left forearm was meant to carve his wife's name in beautiful Hindi script, but I'm sure it's not spelled 'Vihctoria'. As a permanent installation in our skin, tattoos are notoriously hard to remove. In the past, acids, salt abrasion and even surgical excision made the process slow and agonizing. In recent decades, however, laser technology has made the task somewhat easier. Lasers exploit the same immune cells that tried to originally engulf the large pigment particles. The physics of laser removal is remarkable: a laser pulse shoots through the skin and is absorbed by a pigment particle for a few nanoseconds (one nanosecond being 0.0000000001 per cent of a second – an unimaginably short period of time), heating the surface of the pigment to thousands of degrees before the energy of the laser collapses into shock waves, breaking apart the pigment but not burning the surrounding skin. To work, the frequency of the laser light is targeted towards a specific pigment of colour. Blacks and dark blues are the easiest to remove, with yellows and whites lying at the limit of the tattoo emission spectra. Once the particles have been broken up into manageable chunks, macrophages proceed to dine on them and move them away from the skin over the following days.

For the uncommitted among us – those who would like to experiment with a tattoo but are worried about the issue of permanence – revolutionary technologies are now popping up. A group of former students from New York University launched a start-up in 2018 called Ephemeral Tattoos.[13] Their ingenious idea is to utilize our own immune system to attack pigment particles in the same way as traditional tattoos, but with a subtle difference. The ink comes in much smaller droplets than conventional tattoos, which are encased in larger spheres made of a translucent biomaterial. Our macrophages aren't able to eat and remove these spheres, but over a specified period of time, such as a year, the spheres break down, releasing the droplets in manageable chunks that the macrophages can now handle. At this point you can decide to have the tattoo drawn permanently, adjusted – or just leave it to fade away.

New tattoo technology has not been universally welcomed by its practitioners, though. A fledgling tattoo artist I met in London voiced her concerns about the future of her new job on hearing that a group of French students had programmed a 3D printer to draw a perfect circle on a volunteer's forearm. She said that tattoos have always been, and should remain, a creative expression of the artist, not a cookie-cutter marking that's downloaded from the internet and drawn by an automaton. I'm not sure anything will change soon, anyway. Like the intricate initiation ceremonies of the Kaningara people of Papua New Guinea, the process of having a person inscribe meaning on to your skin, even if it is simply 'No Regrets' on your left shoulder, is a sacred, human transaction.

One development that's very exciting (or terrifying, depending on how you look at it) is that material technologies are making it a very realistic probability that huge amounts of information could be collected or stored in our skin. Tattoos that constantly monitor your body temperature and blood alcohol level, as well as ones that can hold personal information and be read as a QR code, already exist,[14] as do tattoos containing carbon electrodes that can read our emotions by detecting electrical signals from facial muscles.[15] These cutaneous computers could even be powered by biobatteries that run on the lactate found in sweat.[16] In 2017 researchers at MIT even developed ink made up of genetically programmed bacterial cells.[17] Their experimental tattoos are 3D-printed on to the skin in the shape of a tree, and individual branches light up in different colours in response to stimuli, whether it be body temperature, pH, or external chemicals or pollutants.

The relentless march of technology makes digitizing our skin seem an increasingly conquerable frontier. If skin is the gatekeeper to our secrets, maybe it's not surprising that this new marriage of personal information with the physical self has prompted concerns of Orwellian state control. The question is not just how we are going to synergize man and machine, but what we do with the technology when it inevitably arrives.

For as long as tattoos have been communicative, they have also been curative. On 19 September 1991, two German hikers were traversing the Austrian–Italian border, high up in the Ötztal Alps. As they began to cross between two mountain passes they stumbled across a stiff, naked body sprawled face-down on the floor. Rushing over to what they thought was an injured hiker, they saw that the lower half of its body was completely frozen and was actually inside the glacier. He had clearly been there for a while. When the frozen corpse was finally prised from the mountainside and analysed by scientists and archaeologists, it was found to be from around 3,300 BC. 'Ötzi the Iceman' is Europe's oldest mummy and is probably the most scientifically analysed human ever to have lived. He is a priceless time capsule, giving us a glimpse into a time before recorded European history. Ötzi clearly led a wild life, and theories abound as to how he came to his gruesome death at the age of around forty-five. In one of the many discoveries by scientists in this episode of Neolithic *CSI*, X-rays have revealed that Ötzi had taken a severe blow to the head, and show a flint arrowhead lodged in his shoulder. Intriguingly, he evidently didn't go down without a fight, and had the blood of four different people on his belongings: one on his coat, another on his dagger and, unbelievably, the blood of two different people on one of his arrowheads. DNA analysis shows that he was at high risk of heart disease, was lactose intolerant and harboured an internal parasite called whipworm.

One of the most surprising findings, however, was that his body is covered in small tattoos. In 2015, multi-spectral imaging analysis revealed that there are sixty-one tattoos in total, mainly consisting of series of horizontal and vertical lines and small crosses, most probably made by rubbing charcoal into punctured skin.[18] Not only are they clearly intentional, but they also appear purposeful as well as beautiful. The locations of the tattoos also suggest that they are not simply cosmetic or cultural. Most of these markings cluster over his ankle, wrist and knee joints, as well as his lower back, areas where he was suffering from arthritis. Other tattoos seem to be placed at acupuncture meridians, with 80 per cent overlapping with traditional Chinese acupuncture points. It appears that the world's oldest tattoos were probably medicinal.

Over 5,000 years after Ötzi the Iceman came to his unfortunate end, there are still a few groups around the world who carry out medicinal tattooing. Lars Krutak, an extraordinary American tattoo anthropologist, has spent time with Yupiget women on St Lawrence Island off the coast of Alaska. They practise 'skin stitching', which is apparently as painful as it sounds. This 'epidermal embroidery', as Krutak calls it, is carried out by women in their eighties and nineties, who stitch a pigmented needle across the skin with the purpose of shutting down potential passageways for evil spirits to enter the body.[19,20] Something close to what Ötzi may have experienced can be found amongst the Kayan people of Borneo, the Southeast Asian island where I spent the first few years of my life. Whenever a Kayan man or woman has a break or sprain, they have a dot tattooed over the joint.[21] A number of tribespeople have numerous dots over one ankle, as the treatment is repeated until the damaged body part is healed. It may be that Ötzi had multiple tattoos in the same spot. The skin is the meeting place of inner disease and outer threats, so no wonder many cultures have used tattoos to both heal the inside and fend off malicious spiritual forces from the outside.

Colin Dale, a tattooist based in Copenhagen, had a client suffering from arthritis, asthma and recurrent headaches. With the help of an acupuncturist he decided to tattoo small dots over the client's body in similar locations to Ötzi's marks. While this didn't completely cure the conditions, all the symptoms improved drastically and were still improved after a year. There's no robust evidence suggesting that marking our outsides can cure our insides, but humankind's propensity to medically tattoo or puncture the skin is intriguing. Current evidence suggests that acupuncture can relieve pain in the short term, although it is not dependent on the location of needle insertion. Whether this effect is brought about by the inflammation and nerve stimulation from piercing the skin, or is simply a placebo, is not yet known.

When it comes to the brain influencing the body with the placebo effect, the more significant the treatment, the greater the effect. If a placebo sugar pill is given to a patient alongside a consultation with a doctor, the patient is likely to feel better than if she had taken a sugar pill alone. Likewise, a bigger pill has a greater effect than a smaller one, and an injection produces better results than a tablet. It's perhaps not surprising that the invasive, intimate and time-consuming interaction between acupuncturist and patient helps the mind and, probably, the body. And it's not too much of a stretch to argue that tattooing could display a similar effect. The process of tattooing, although painful, releases adrenaline and endorphins. Evidence also suggests that tattooing creates a positive self-image and confidence that certainly lasts for weeks, and in some cases, throughout a lifetime. Research from the University of Alabama found that the inflammation, pain and stress of having a tattoo etched into your skin temporarily lowers your immune defence and has also been shown to make you more likely to catch a cold.[22] But intriguingly the research also found that repeated tattooing actually bolsters the immune system, making you better at fighting off common infections. If on the first day of your gym membership you head over to lift the heaviest weight, it will put enormous stress on your body; but after repeated training it becomes lighter, and you become stronger. So, too, with tattooing. The tattooist's vibrating needle packs another punch, too: its immune stimulation has also been shown to have a potent 'adjuvant' effect. Adjuvants are molecules designed to boost your immune response; when you have a vaccine injected into your arm, it contains an adjuvant. Research conducted at the University of Heidelberg in Germany found that the act of tattooing is actually more effective at generating an immune response to DNA vaccine than some conventional adjuvant molecules.[23]

When it comes to permanent health benefits, careful tattooing may also come to the aid of visible skin conditions. Tattoos can permanently camouflage scars and the depigmented patches of vitiligo, as well as creating the illusion of short hair for those with hair loss. The transformative power of a tattoo is no more apparent than in a medical tattoo following a mastectomy for breast cancer. Increasingly, these range from tattooing a replacement areola to give the breast a normal appearance, to the opposite: having bold, life-affirming tattoos placed across the mastectomy scar. A tattoo is not always welcome, however, particularly if it reminds a person of disease and death. A number of women who undergo radiotherapy for breast cancer have dots tattooed on their skin as a target for radiotherapy beams, and many resent the constant reminder on their skin of cancer. However, a team at the Royal Marsden Hospital in London found an ingenious way to give women control over the medical pigmentation.[24] They conducted a study in which half the women undergoing radiotherapy had conventional tattoos to guide the beams, whereas the other patients had fluorescent tattoos. These emerged during the rave culture of the 1990s; their special fluorescent ink means the tattoos are invisible in normal circumstances but light up under ultraviolet light. The women with these 'invisible tattoos' had markedly improved body confidence as a result, reporting that they felt more in control of their own body than if they had conventional markings.

The social skin is also where medicine and messaging meet; in the form of 'medical alert' tattoos. I have seen a number of diabetic patients sporting markings on their forearm or wrist in case a diabetic coma steals verbal communication from them. While medical alert tattoos can be helpful, they need to be approached with caution. During the Cold War, the US government considered tattooing blood types on to citizens, providing millions of walking blood banks in the event of a nuclear strike.[25] This programme was not rolled out, despite two short-lived trials in Utah and Indiana, because ultimately doctors did not trust tattoos with life-or-death decisions. In 2017, doctors in Miami University Hospital were assessing a seventy-year-old unconscious patient with a high blood-alcohol level who was rapidly deteriorating.[26] Unbuttoning the man's shirt to place ECG leads, the medical team was immediately greeted by the words 'DO NOT RESUSCITATE' emblazoned in green ink across his chest, with a blurry, faded signature below. Unable to contact his next of kin or find any official documentation, the doctors were hurled into an ethical dilemma. They eventually decided to respect his presumed wishes and he died overnight. While it may look like a clear and permanent way of demonstrating one's wishes, our minds change faster than our skin, so what if the man had wished to be resuscitated but hadn't had the time, energy or finances to endure the arduous process of tattoo removal? What if he was inked as a joke or a drunken bet? These are a few of the reasons why some jurisdictions, such as the UK, maintain that you can only request DNAR (Do Not Attempt Resuscitation) by signing a written form countersigned by a witness. As with any form of human communication, our skin is also capable of miscommunication and mistrust.

Tattoos have also been used to prepare for the end of life. When European crusaders set sail for the Holy Land during the Middle Ages, some had large ink crosses tattooed across their chests so that if they died they could be identified and given a Christian burial. The late tattoo artist Jesse Mays continued this tradition in a studio near Camp Lejeune, a marine base in North Carolina. During the wars in Iraq and Afghanistan, Mays would have soldiers of all ranks coming in for 'meat tags'. This permanent dog tag ID was tattooed into their skin above their ribcage, stating the soldier's name, religion, blood type and medical conditions, such as diabetes.

There are even tales of the ashes of the departed being made into ink, to live on in the physical identity of a loved one. These spiritual tattoos make us feel more alive, and see us through to death.

Why do humans tattoo? On the one hand there are the institutionalized tattoos of civilizations – from the Maori in the Pacific to the Hamar tribe in Ethiopia – which are all about social cohesion and communication. On the other hand, the modern, 'Western' tattoo is a sign of individuality and rebellion. In the old world, Christianity, Islam and Judaism had effectively banned skin-marking for centuries. But from the clash of these separate skin ideologies on the beaches of the Pacific, with the arrival in New Zealand of James Cook and the *Endeavour* in 1769, we see that human reasons for communicating through the skin are more similar than they first appear.

HMS *Endeavour* set sail from Plymouth on its voyage of scientific discovery carrying the little-known naturalist and botanist

Joseph Banks. This flamboyant, aristocratic Old Etonian was there to study plants, but his fascinating (albeit sloppily written) journal largely ended up documenting the people the expedition encountered. Among the many discoveries Banks recorded when the *Endeavour* laid anchor on the island of Tahiti (his journal includes the first written description of surfing), he watched with astonishment the process of skin-marking and penned the word 'tattowing' for the first time. When you say this onomatopoeic word, which is derived from the Polynesian *tatau*, you can feel the tattoo artist striking the wooden comb, often impregnated with shark teeth, into the skin of the islander. Banks noticed that while all of the islanders tattooed, the practice was just as much about individuality as conformity.

> 'Every one is thus marked in different parts of his body, according maybe to his humour, or different circumstances of his life.'[27]

It didn't take long for inquisitive European sailors to want to adopt these practices for themselves. The inkings quickly developed meaning, and achievements and stories were proudly emblazoned on their outer covering: an anchor for sailing the Atlantic, a turtle for crossing the equator and a swallow for completing journeys totalling 5,000 nautical miles. Europeans were also quick to adopt the spiritual, superstitious quality of skin-marking: a pig on one foot and a rooster on the other would protect you from drowning and the letters 'HOLD FAST' across both sets of knuckles would help you cling to the rigging during a storm. A few centuries later, in a clinic in Birmingham – the most tattooed city in the UK (itself one of the world's most highly inked countries) – I examined a patient who embodied the practice of modern tattooing. A middle-aged social worker, he sported both an anchor and a swallow, his shoulders were emblazoned with coloured images of lions, his daughter's name was written above his heart and most other areas of his skin were covered in a miscellany of Celtic crosses and Chinese characters. At first glance, modern tattooing seems a complete freedom of choice, and in a sense it is. But studies also show that people choose Western tattoo designs – often seen as symbols of uniqueness and personality – based on popularity instead of individuality, confirming that tattoos are also about conformity. Johnny Depp's words 'My body is my journal, and my tattoos are my story' echo the Maori *Tā moko*, which celebrate individual achievement and experience, but also look back to tribe and ancestry. It's a story about ourselves, but one that we want others to see and engage with.

Our most human organ is paradoxically our most social by being our most individual. The fact that humans consciously create a physical manifestation of symbols and ideas on their bodies is utterly remarkable. When we permanently put meaning on our skin it takes on incredible power, stating who we are and who we want to be. I once met a Naga 'tiger warrior' in Northeast India who argued that his tattoos were his only true possessions, as they were the only things he had created that went with him to the afterlife. Where does the tattoo end and the person start? By altering our appearance, we attempt to transcend our natural bodies in some way. In a homogenized world where it's hard to stand out with clothing and cosmetics, tattoos are a way of putting our ideal inner self on our outside.

9

The Skin That Separates

The sinister side of the social organ: illness, race and sex.

'I would rather not see than be seen like this.'

A South Sudanese man with onchocerciasis (river blindness). This disease damages both the skin and the eyes.

The fan had broken in my clinic. As I wasn't allowed to remove my ill-fitting white coat in the heat of a Tanzanian hospital, it was the worst-case scenario for a cold-loving Brit on his first day in Africa. The small room was empty, save for a cork noticeboard smattered with drug charts and HIV education leaflets. Sitting with me was Albert, a local doctor who was my teacher and translator. At the opposite end of the desk was Dani. Head bowed, eyes fixed on his shoes, he was my final patient of the day. His stature and facial features were similar to those of many young Tanzanian men, but it was clear that he was an albino. His white skin had a delicate, almost translucent appearance and his head was crowned with straw-coloured hair. Albinism is caused by genetic mutations, and Tanzania has the highest incidence of this condition in the world. The mutations disable the production of the skin's black pigment, melanin. Lacking this protective wall paint, each albino is consigned to a life of perpetual sun avoidance and recurrent skin cancer. I reached for my dermatoscope and scanned Dani's snow-white skin for signs of cancer. If I found anything, I could either blast away at it with liquid nitrogen or refer him for surgery. Dani responded to my questions about his previous cancers with diminishing interest, and as the consultation went on it became apparent that the physical aspect of Dani's disease was the least of his worries. As his story slowly unravelled, I discovered that although the sun was a torment to him, it came a distant second to a fear of his fellow man.

Dani had been rescued from his village as a child after his uncle had tried to kidnap and kill him. Since then he had spent his life in an isolated, high-walled school built to protect albino children from their own people. Now leaving the relative safety of the school, he felt little prepared for an utterly hostile world. Albinos in Tanzania have long been referred to as *zeru* ('ghosts' in Swahili) or *nguruwe* ('pigs'), but the scale of the murder and maiming of these people is relatively new. The greed of witch doctors and the poverty of the rural population have contributed to a belief that the body parts of albinos bring good fortune, wealth and political power. Other beliefs include albinos being demonic spirits, the ghosts of European colonizers or the progeny of a woman's unfaithfulness with a white man. The crushed limbs of albino children are said to cure any ailment and demand the highest price. When a full set of albino body parts can fetch up to $100,000, it is easy to see why witch doctors are not short of recruits with murderous intent.[1]

The cruel irony is that albinos, because of their melanin deficiency, already have a low life expectancy. Dani told me that it was worse for girls and women, with some rural Tanzanians believing that having sex with an albino cures AIDS. Now a young man, he said he no longer feared for his life, but he had the look of someone resigned to the existence of an outsider. The plight of albinos in East Africa is not history; it is a silent but growing humanitarian crisis. Rough approximations suggest the number of albinos abducted and killed since 2000 to be in the low hundreds; an African doctor I worked with at an albinism specialist centre is certain that actual numbers are much higher, the secret slaughter being kept behind closed doors and within families.

Skin is a physical material, as real as our heart and liver, but it is simultaneously – and uniquely – a social substance. A single genetic mutation, affecting solely the melanin production in skin, can ruin someone's life, and even end it at the hands of others. Albinism in East Africa is a terrifyingly visual insight into how, even when culture and ethnicity are accounted for, skin appearance is an easy vehicle to define someone as 'other', to stoke fear and satisfy greed. Closer to home, I've seen that for the skin to divide people it doesn't have to be too dark or too light, but too *different.* I recall a young Pakistani woman from Birmingham whose face was patchy-white from vitiligo. A few minutes into discussing her many past attempts at treatment she burst into tears, describing in fits and starts how she would never be able to marry. A few months later I met an equally despondent Indian woman whose face had been darkened due to a condition called melasma. She had very dark brown blotches spread symmetrically across her face. This darkening is caused by oestrogen and progesterone (both of which are produced in abundance during pregnancy, although this patient was not pregnant) stimulating melanocytes to pump out melanin. For this reason melasma is often termed 'the mask of pregnancy' and it is hypothesized that during pregnancy the body is trying to protect folic acid in the skin from sun damage. This could explain why generally women have their darkest skin colour during their reproductive years. Of these two patients, each from a base of light brown skin, one had lightened and the other had darkened, but the social consequences were the same.

'Black lives matter'. 'Redskins controversy'. 'Hollywood whitewashing'. Each phrase appeared in a headline of a major American newspaper within a single week in 2018. More than ever, debate and discourse is loaded with colour. Diseases affecting pigmentation ostracize individuals from their society, but it is the completely natural differences in skin colour that have caused the most widespread division between human beings throughout history. How can something as seemingly small as the concentration of melanin in our 1mm-thick outer covering be the cause of so much pain and suffering? Skin colour is largely determined by the type and concentration of melanin in the skin and, as we saw in Chapter 3, this is due to the skin's role as both a fortress and a factory. The octopus-like melanocytes pump out the dark pigment melanin to protect us from UVB rays, yet our skin also opens itself up as a chopping board, eager for these rays to dice up precursor molecules into active vitamin D. As humans started migrating from the hot, sunny regions of Africa and the Middle East, their skin began to walk a tightrope: too much melanin in a region with limited sunlight would cause vitamin D deficiency; too little in a sunny climate would allow sunlight to wreak havoc with skin's DNA, as well as reducing our body's levels of folic acid, necessary for producing healthy progeny.[2] Over thousands of years of migration and adaptation, populations that moved further from the equator into areas with lower UV light began to develop lighter skin. A world map showing the indigenous distribution of human pigmentation almost perfectly maps on to NASA satellite imaging of the concentration of UV exposure striking the planet. There are notable exceptions, which actually help to strengthen the theory: the dark-skinned Inuit live far away from the equator, but this is most probably because the unusually high levels of vitamin D in their diet of fish and whale blubber compensate for an inadequate intake by the skin. It is also likely that their pigmented skin protects them from extremely prolonged (and snow-intensified) dosages of UV light in the summer months.

Skin's fine-tuning of melanin levels during these migrations has given the human race a wonderful and unique rainbow of colour. The sheer vastness of the range of human skin colours is a result of a number of different genes regulating different types of melanin. Individuals with light skin are likely to produce more of the red-yellow 'phaeomelanin' (which also gives lips, nipples and red hair their appearance) and those with darker skin produce more of the brown-black 'eumelanin', which is by far the most abundant pigment in human skin. Melanocytes are covered in tiny molecules called melanocortin 1 receptors (MC1R) and when these are activated they reduce the cell's phaeomelanin production and replace it with eumelanin. In most people with red hair, white skin and freckles, the MC1R gene is mutated and the receptor doesn't work. This mutation provided a benefit to humans who had migrated to Northern Europe, with its low levels of UV light, and it is still very common today, particularly in those of Celtic origin.

But even our adaptable skin isn't quick enough to adjust to globalization. It is possible today to cover distances in a few hours that human skin took millennia to adapt to. Light-skinned Europeans who have in recent history moved to a region with high UV exposure (such as Australia), or who regularly visit sunny countries, have a significantly increased risk of developing skin cancer. Conversely, darkly pigmented migrants who move to northern latitudes are likely to suffer from the osteoporosis, muscle weakness and depression that come with vitamin D deficiency. Possibly the most well known of these migrations was the forcible relocation of some twelve million Africans to North America during the transatlantic slave trade. As our most social organ, skin also shows humanity at its worst, and it wears the scars of history. Skin is a perimeter fence that delineates our inner being from the outside world, defining ourselves and keeping others out. It's also our most visible organ. This has made skin a social weapon, exploited by two forces that plague the human condition: namely, the search for identity and the desire for power.

Who am I? What is my purpose? Where do I fit into the world? One of the fundamental ways of recognizing that we exist at all is that we perceive other things – other people – and how they respond to us. To define our 'self' is to define the 'other'. German philosophers, including Hegel and Husserl, dedicated their lives to understanding, and attempting to explain the relationship between, our consciousness and our perception of the outside world. Their approach led to the concept of 'othering' – the process of defining others as different from us in our development of self-awareness. This is equally true for groups. It is immeasurably easier to process information in discrete categories than to manage the complexity of reality. We therefore develop concepts of ourselves and, through jokes and insults, we easily and negatively stereotype the 'other'.

The Polish sociologist Zygmunt Bauman argued that these group identities also form binary, opposing categories – animal and human, stranger and native, them and us – a conclusion almost certainly informed by his Jewish family's experience of Nazi genocide.[3] Animosity between different tribes and nations is a characteristic of human interaction dating from the earliest times, from Jews distancing themselves from Gentiles and Greeks pitting themselves against Barbarians. But 'colourism' – racism in its skin-focused sense – accelerated during the sixteenth and seventeenth centuries. The European Age of Exploration, with its nascent empires across the globe and the emergence of the slave trade, was supported by pseudoscience and taxonomy that sought to justify racism. The field of physiognomy (literally, 'judging nature', and now completely discredited) claimed that measuring physical characteristics can reveal inner character. Ironically it is the modern study of skin-colour genetics that supports the prevailing scientific and anthropological consensus that all humans are, biologically, one single race.[4] A 2017 study found that the numerous gene variants that lighten or darken skin colour, eight of which the study measured, are spread across the world and shared unevenly across different ancestral groups, explaining, for example, the diverse range of skin colours among African ethnic groups.[5] The back-and-forth migration of humans has intermixed these genes, and the palette of human colour varies widely within geographical regions and ethnic groups, making skin a poor marker for ancestry, let alone the unsupported idea of biological race. When South Africa came out of apartheid (literally, 'apart-hood') and held its first democratic elections in 1994, Desmond Tutu described the country as a 'rainbow nation'. The biological reality of our physical skin informs our social skin. It echoes the Archbishop Emeritus, imploring us to celebrate our individual genetic variation while humanity remains as one.

The skin's power to separate spreads much further than pigmentation. In medieval Europe, skin disease at both ends of society was caricatured, from the nutrient-deficient poor at one end to the coarse, ruddy skin of an overindulgent elite at the other. In the late eighteenth century, when the emerging industrial cities began to fill, skin soon became a middle-class battleground, a marker of health and social standing. Medical historian Richard Barnett argues that 'high collars and long skirts hid more than bourgeois modesty' during the industrial revolution. They hid external signs that supposedly revealed internal flaws. 'The itch', which in most cases was probably scabies – a ubiquitous condition in eighteenth-century Britain – was seen as a sign of both poverty and moral failure.[6] This isn't just history; I remember the response of a middle-aged and middle-class patient when I diagnosed her with scabies: 'It's *so* embarrassing – I'm not supposed to get this.'

While the skin as a canvas of cleanliness, delineating class, is a relatively recent social development (roughly three centuries old in the West), human skin has always had a primal power to produce a fear of contagion. None more so than one of humanity's oldest foes: smallpox. Smallpox is caused by an unassuming, brick-shaped virus called *Variola* and it has been responsible for an unimaginable number of deaths in human history. First, tiny red spots would appear on the tongue, accompanied by a fever, a splitting headache and nausea. Within twenty-four hours the rash would cover the whole body and the flat marks transmuted into fluid-filled bumps with characteristic 'umbilicated' depressions in the middle. During the week-long period from the onset of the eruption to the scabbing-over of the bumps, the sufferer could spread death with a simple touch.

When Spanish settlers brought smallpox from Europe into an unprepared New World, in some areas it wiped out 90 per cent of the Native American population, far more than famine and war combined. The many who were infected by smallpox but did not die were left with permanent, disfiguring scars from the distinctive pitted blisters of the disease. Smallpox did not discriminate – or so it seemed. In 1796, a time when the ravages of smallpox affected every community in Europe, a dandy and maverick of a doctor, Edward Jenner, working deep in the English countryside in Gloucestershire, noticed an interesting anomaly. It was common country folklore that milkmaids had fair, unblemished complexions. As he strolled down the country lanes, passing by farms, fields and villages, Jenner realized that milkmaids were the only people whose skin was not scarred by the smallpox rash. Their skin was separate. After much deliberation he hypothesized that they contracted cowpox, the milder bovine equivalent, which conferred immunity to smallpox.[7] Testing his theory, he took pus from Sarah Nelmes, a milkmaid with cowpox, which he injected into a cut in the arm of a local village boy, James Phipps. A few days later, Jenner then injected the boy with the scab material of someone infected with smallpox. Nothing happened. This eureka moment, on the back of 'fair skin' folklore, led to the world's first vaccine (*vaccus* being the Latin for cow).[8] With this discovery Jenner has arguably saved more lives than any other scientist in history. But it took time to slay the speckled monster of smallpox. It's likely that it killed around 400 million people in the twentieth century alone. Its last victim, in 1978, was the medical photographer Janet Parker. Working in Birmingham Medical School's anatomy department (and, incidentally, in the room where I was taught anatomy as a student), she was accidently exposed to the virus, which was being grown and studied in the laboratory a floor below. The incident led to the destruction of all worldwide stockpiles, except in one laboratory in the USA and one in Russia. It's no surprise that the remaining vials of this red plague fuel rumour and fear of biological war.

Smallpox terrifies because it is contagious and often fatal, but many skin diseases invoke fear not by being deadly, but by being deforming. Two years after achieving independence in 2011, South Sudan descended into civil war and ethnic violence. In East Africa I met a young doctor named Elijah, who had fled the world's newest country as a refugee, to discuss a terrifying infectious skin disease endemic in rural areas. 'If the ethnic struggles divide villages,' he said, 'onchocerciasis divides families.'

Onchocerciasis, or 'river blindness', is caused – as we saw in Chapter 2 – by a parasitic worm that lives in the saliva of blackflies. In humans, as well as causing preventable blindness, the parasite produces an intensely itchy and disfiguring skin disease. Elijah said that his patients often found the skin condition to be worse than blindness, and had once overheard the comment 'I would rather not see than be seen like this.' The indescribable itch of onchocerciasis causes sufferers to scrape their skin until its physical change causes social disfigurement. In addition, severe spotted areas of depigmentation result in what locals call 'leopard skin'; atrophic, loose skin is called 'lizard skin'; and thickened areas 'elephant skin'. These terms are far from trivial. In remote areas, having this animal-like appearance is a curse and individuals risk being thrown out of families and communities because of it. The attribution of animal-like qualities to humans is called zoomorphism, but when it concerns dehumanizing people, 'beastification' might be more appropriate. History has often shown the dehumanization of groups through their skin in the path to their elimination. Jews, with their skin supposedly infected by *Judenkrätze*, the mythical 'Jewish itch', and with exaggerated facial features on Nazi posters, were portrayed as migrating, wandering rats to Germans in the 1930s. 'Rats are the vermin of the animal kingdom, Jews are the vermin of the human race,' declares the narrator of the film *Der*

Ewige Jude, compulsory viewing for the SS soldiers heading east to carry out the inhuman acts of the Final Solution.

Many societies throughout history have projected separate ideals for men and women, and the skin is a strong, but overlooked, dividing line used to define the sexes. In many cultures it is considered feminine to have comparably lighter and more delicate skin, a translucency that implies openness, innocence and sincerity. The male skin, on the other hand, has most often aspired to represent swarthy, impenetrable armour. Although these differences are clearly a projection of a society's values on to skin, the intriguing question is whether it has a biological inspiration. In any given racial group, females generally tend to have lighter skin. This is because women are likely to have high vitamin D and calcium requirements for childbearing. Male skin is approximately 25 per cent thicker as a result of higher levels of testosterone, and the thickness in the uppermost layers of the epidermis makes it noticeably rougher. Men also tend to have a higher density of collagen in the dermis, and lose it at a slower rate than women as they get older. This begs the question why male skin doesn't seem to age at a slower rate than women's. While there's no definitive answer, it's been shown that, on average, men tend to open themselves up to more unprotected sun exposure during their lifetime, probably cancelling out their ageing advantage.[9]

Male warriors of legend have extended their skin's thickness from the biological to the metaphysical, literally wearing their skin as impenetrable armour. The Greek hero Achilles was dipped in the River Styx as a child and the fabled German warrior Siegfried, rather more dramatically, was bathed in dragon's blood. These baptisms bestowed on the mythical figures skin that no weapon could pierce. But tiny spots of vulnerability, on Achilles's undipped heel and a small area of skin between Siegfried's shoulder blades, would prove to be the downfall of these warriors. Even for seemingly invincible heroes, skin is our most vulnerable, and most human, organ.

The gender separation of the skin has, unsurprisingly, been propagated, exaggerated and exploited by folklore and questionable science throughout history. Daniel Turner's 1714 *De Morbis Cutaneis: A Treatise of Diseases Incident to the Skin* is arguably the first British dermatological textbook. In it, Dr Turner argued that a pregnant woman's 'imagination' could create markings on the skin of her foetus.[10] This theory of 'maternal impression' (now completely debunked) was a reflection of the widely held view that if an expectant mother saw something that gave her a sudden terror, a projection of the object would be imprinted through her emotions on to her unborn child. Thankfully, modern genetics has confirmed that the hairy mole on your back didn't come from your pregnant mother being chased by a bear. This history still lingers, however, with the German and Dutch words for 'mole' – *Muttermal* and *moedervlekken* respectively – both translating as 'mother-spots'.

So what, actually, are moles? And what about other birthmarks? Available in a range of sizes and a catalogue of colours, they are all benign overgrowths of different components of the skin. These are usually either pigmented (caused by melanocytes) or vascular (originating from blood vessels). The common 'mole' (an easier term than its official 'common melanocytic nevus') is anywhere from dark-brown to black in colour, and is caused by small, localized genetic mutations in the foetus between the fifth and twenty-fifth weeks of gestation – the earlier in development these proliferations occur, the larger the mole. Moles are with us for life, but that's not the case for all congenital birthmarks. 'Mongolian spots' are flat, bluish patches usually located on the back and buttocks of infants, and almost always disappear by puberty. Their name was coined by the German doctor Erwin Bälz, who served as physician to the Japanese Imperial Family at the end of the nineteenth century. He incorrectly thought that they were largely restricted to his Mongolian patients, whereas these birthmarks are common in populations across the whole of Asia, Oceania and

Latin America. They are caused by the migrating melanocytes failing to complete their journey to the epidermis during embryonic development, instead getting stuck in the lower half of the dermis. We see them almost translucently behind the layers of the epidermis, giving these spots their curious tinge of blue. Mongolian spots are classed as 'macules', meaning there is a change of colour without any surface elevation or depression. Other macules include café-au-lait spots, appropriately milky-coffee coloured. These are themselves harmless but are often the heralds of a whole host of genetic disorders – such as neurofibromatosis, in which tumours grow along nerves.

When it comes to vascular birthmarks, I have rarely visited a neonatal ward without seeing a distinctive pink mark on the back of an infant's neck. The 'nevus flammeus nuchae' – or, as midwives call it, the poetically colloquial 'stork bite' – is common on white skin and is usually temporary. Another common benign vascular growth of infancy is the infantile haemangioma, or strawberry mark. Raised, vivid in colour and sometimes large, these can look alarming but the majority disappear, leaving no visible marks, and the cause of their rise and fall remains largely a mystery to scientists.

Some visible birthmarks, of course, do not vanish and, depending on their location and their wearer, can have crippling mental and social consequences. It is believed that the distinctive 'port-wine stain' is caused by a localized lack of nerves for controlling blood-vessel dilation and constriction, leading to permanent dilation and a pooling of blood. Not only have women historically been blamed for giving these whorls and patches to their offspring, but they have also been judged for their own, although the wide range of interpretations, from impurity to individuality, says a lot more about the viewer than the wearer. Compare the Salem witch trials – where women were marked out as witches and executed in part for skin markings that were interpreted as associating them with the devil – with the enigmatic rise of the beauty spot in the eighteenth century.[11] Perhaps beauty spots highlighted the paleness of the skin of the wearer, or covered the pockmarks of smallpox. Perhaps the allure was more mysterious. Karen Hearn, art historian at University College London, argues that the positive viewing of these spots harks back to ancient times: 'the goddess Venus was said to have had a mole. This slight defect on an otherwise perfect body made her seem even more beautiful.'[12] Some ancient cultures took inferring character from our skin markings to extremes with 'moleosophy', a forgotten but equally unscientific cousin of palm-reading.

The skin is our largest sexual organ, and in society's eagerness to judge and separate, it has frequently been used to scandalize. The Italians called it 'The French Disease', the French 'The Italian Disease', the Russians 'The Polish Disease' and the Turks 'The Christian Disease'. It was 1495, and the French army was besieging the Italian city of Naples. As if from nowhere, French troops and their Spanish mercenaries began to develop bulbous pustules that would seep with rancid pus before the skin itself began to fall off. 'The great pox', most commonly known as syphilis, had arrived in Europe. One theory is that it originated in the Americas and travelled on European vessels coming back from the New World following Christopher Columbus's 1492 voyage. In the early years of the Columbian Exchange, where goods, ideas and diseases were traded across the Atlantic, the arrival in Europe of syphilis was a softer, if ironic, blow than the New World receiving the deceptively named but deadly smallpox. In Europe it soon became clear that syphilis was associated with sex, and severe stigma quickly followed suit.

This extremely social disease makes a distinctive and relatively predictable journey across the skin, but its medical progression has nonetheless fascinated doctors for centuries. This ugly disease is caused by the beautiful, spiral-shaped bacterium *Treponema pallidum*, a form of bacteria termed 'spirochetes'. They look like coiled snakes or, if you are as gastronomically inclined as one of my tutors, who managed to make any bacterium sound edible, like curly fries. They are obligate parasites, meaning that they can only survive within the host, and they travel to new hosts through either sexual contact or direct contact with an open skin lesion.

Just suppose that a couple of these serpentine bacteria depart from their home in the vagina of an infected female to form a new colony on the penis of an uninfected man. In the few weeks following sex, the spirochetes start to make the point of contact on the penis their home, destroying the tissue and forming a small ulcer, called a chancre. Although painless, the chancre is a snake pit, a crater brimming with fluid nursing a growing population. This initial sign, the 'X marking the spot' of the supposed transgression, will usually disappear within one or two months. Although this painless 'primary syphilis' can be hidden from public view, the skeletons will come out of the closet a few months later. The spirochetes leave their nest on the tip of the penis and journey through the new host's lymphatic system, eventually reaching the bloodstream. On visiting the skin, they inflame the lining of the dermal blood vessels and cause the generalized rash of secondary syphilis: red, non-itchy, macular (flat) and papular (raised) spots appear on the trunk and spread across the limbs to the host's palms. Many a supposedly celibate priest or virtuous woman (and even a pope) would have had their secrets blurted out on the surface of their body. The unambiguous marks were there for all to see.

Then everything goes quiet. Symptoms disappear. The disease enters a latent phase, where the spirochetes retreat to the small blood vessels of the internal organs, going completely unnoticed for anywhere between two and twenty years. The final, and often fatal, stage – of 'tertiary syphilis' – is now rarely seen in the developed world, following the advent of antibiotics. Even with low levels of spirochetes residing in the host, his immune system goes into overdrive, forming gummas – balls of inflammation with a core of immune cells and a thick coating of fibroblasts. In the pre-antibiotic age, these growing gummas would go on to damage any tissue inside the body, and indeed deform the skin and disfigure the face, delivering a slow, humiliating death. Up until the twentieth century, rubbing mercury on to the skin or inhaling it in vaporized form was the treatment of choice (despite being highly toxic with little or no symptomatic benefit) for those who could afford it, leading to the old saying: 'One night with Venus, a lifetime with Mercury.'

If many in society considered that one venereal sin would lead to eternal torment, hiding the skin symptoms from others was as important as treating it. The advent of syphilis in Europe was the impetus for the increase in women (and men) wearing increasingly impenetrable make-up. Meanwhile, Cesare Borgia, a power-hungry Italian nobleman of the fifteenth century and once described as the 'most handsome man in Italy', spent his final years with half of his face obscured by a leather mask, hiding the fruits of his perceived iniquities.

Diseases associated with sexual immorality have throughout history turned skin into a weapon, one that can wreak terrible injustice. In 1932 the US Public Health Service and the Tuskegee

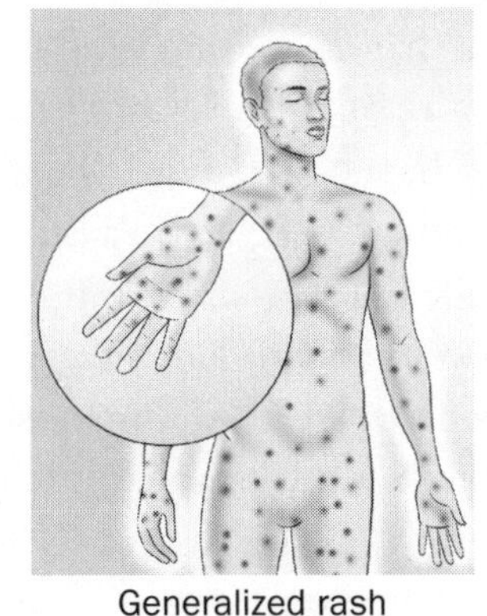

THE STAGES OF SYPHILIS

Institute in Alabama began an experiment to observe the progression of untreated syphilis. Over a period of forty years, almost four hundred black men infected with syphilis were observed, with the reward that they would be given free medical care by the government. But this was laced with a lie: despite penicillin – which could cure this awful disease – being effectively tested in humans in the 1940s, these syphilis sufferers were knowingly not treated with the necessary drug. Twenty-eight men were left to die of the disease and around a hundred more died of related complications. This was able to happen because the black skin of the participants had rendered them powerless in society and the sexual stigma of syphilis allowed the men to be dehumanized even further, to the level of laboratory rats. This dark hour in the history of American medical research led to the formation of the country's National Research Act in 1974, which finally enshrined regulation of human experimentation in law.

With the advent of antibiotics in the first half of the twentieth century, syphilis became cheaply treatable and its scandalous image drifted out of the public consciousness. A few decades later, however, and seemingly out of nowhere, there came another sexually transmitted disease that flashed up on the skin, again revealing not only physical infirmity but also society's interpretation of deviance, scandal and shame. In the 1993 film *Philadelphia,* Tom Hanks plays Andrew Beckett, a senior associate in a large law firm. Early on in the movie, one of his colleagues notices a single purple mark on Andrew's forehead. This seemingly harmless violaceous lump turns out to be Kaposi sarcoma: a rare, AIDS-defining skin disease. It is also a hole that opens on to Andrew's soul as it releases his diagnosis into a hostile society. Andrew hurriedly tries to hide his skin's announcement of what society then regarded as his moral transgressions. In an age terrified of this largely unknown and untreatable disease, and disgusted by its association with homosexual sex, Andrew is sacked.

Through the prism of the skin, and the fear of the disease that lay beneath it, *Philadelphia* was one of the first Hollywood films to overtly address homosexuality within American society. When the 'acquired immune deficiency disorder' (AIDS) epidemic – caused by the human immunodeficiency virus (HIV) – was unleashed upon the world in the early 1980s, the devastating disease was briefly named gay-related immunodeficiency (GRID) due to its explosion in male gay communities in California. Although it was changed to AIDS within a year, the connotations of what was, at the time, widely seen as sexual deviancy have been hard to shake off. AIDS is of course not solely a skin disease, but around 90 per cent of HIV-infected individuals experience skin conditions throughout the course of the disease, and it is often these that reveal the diagnosis to others. The HIV virus wipes out critical components of the immune system, opening the floodgates to a series of opportunistic skin infections as varied and enigmatic as the disease ravaging the body beneath. These can be distinctive, from the fleshy, dimpled bumps of the *Molluscum contagiosum* virus to the herpes virus found in the red-purple cancers of Kaposi sarcoma; but they can also be hard to pin down, with the total war raging in the body causing all manner of cutaneous collateral damage – eczema, shingles, seborrheic dermatitis, scabies, photosensitivity, warts, thrush . . . the list goes on.

When Singapore cracked down on heroin-trafficking from the 'Golden Triangle' (the million square kilometres of poppy fields across Myanmar, Thailand and Laos) in the late twentieth century, new illicit trade routes opened up across the Indian–Burmese border. This incredibly remote region includes the Indian state of Nagaland, and during a visit I made to its mountainside towns and villages, a local doctor summarized the mountain of stigma that faces AIDS patients today. 'It used to be homosexual sin. Then it was the promiscuous truckers and prostitutes along the mountain passes. Now it's heroin injections. "Sex, drugs, rock and roll," as you English say. Some think that if you have the mark of AIDS, you don't belong here, so people who have the disease need to have a hidden life. We can treat AIDS now but people don't come forward and only get discovered when it's too late, or when they've infected someone else.'

Even those who once stood at the top of local government, and high up in the rigid Indian caste system, would be rendered one of the 'untouchables' following rumours of HIV infection. It was inevitably those with a visible, skin manifestation of the disease who were picked out first. In some cases they were considered literally untouchable; I came across a medical professional who believed that HIV was spread by skin-to-skin contact. People did not want to know about AIDS and, by extension, if you had the disease it was better that you did not exist.

The moral panic that followed the outbreak of AIDS helped societies across the globe throw everything at treating it, but paradoxically this drove the 'shameful' condition underground, accelerating its spread through continued ignorance. But all is far from lost for sufferers today. Drugs that target the HIV virus are now cheap as well as effective, but the only way to truly tackle this terrible disease is to treat societal stigma. Coordinated programmes to look after those affected, encourage testing and get the people of Nagaland talking about HIV and AIDS are really working. I have been privileged to spend time in centres for infected children, where I learned that over half of children with HIV have a visible skin condition. Just as important as the anti-retroviral treatment is to give children confidence and hope so that they are not forced to live lives blemished with perceived marks of deviancy.

India is also home to half of the world's sufferers of an ancient, legendary skin disease. Throughout history, leprosy is the condition that best illustrates skin's sway over society. During a visit to Africa, I went in search of the disease's physical and social symptoms. Returning from my trip exploring Maasai medicine on the Serengeti, I hoped to visit a large leprosy centre that I had been told was nearby. With only the name of the centre but no address, I asked around fruitlessly for its exact location until I chanced upon a local doctor, who tentatively gave me the name of a village.

The usual mode of transport to a place of this size is the notorious *dala dala*. This dilapidated, smoky, overcrowded minibus is ubiquitous in Tanzania, its name a corruption of 'dollar'. The *dala dala* epitomizes the real danger of Tanzanian roads, something that would be confirmed by the local orthopaedic surgeons I would later meet. Wedged between a large bag of rice and two elderly Tanzanian *mamas*, who took great joy in my pidgin Swahili, I waited three hours until the conductor, precariously positioned half in, half out of the minibus, slapped on the roof of the vehicle and shouted the name of the village: 'Maji ya Chai!' Its literal meaning, 'tea water', derives from the red-coloured mountain stream that runs through the centre of the sleepy settlement. Even in the village, however, nobody knew the whereabouts of the leprosy centre, despite it apparently being one of the largest in the country. My words of 'leprosy' and its Swahili equivalent '*ukoma*' to every butcher, baker and *ugali* porridge-maker in Maji ya Chai drew blank expressions until one boy, no older than fifteen, motioned me to the back of his motorbike. And so I found myself clinging on for dear life as we bumped across a dirt track, away from civilization and deep into farmland, avoiding potholes and clucking hens. There, quite literally in the middle of nowhere, four friendly nuns, who lovingly looked after the centre and its thirty or so lepers, invited me to visit the inpatients.

With Sister Christie translating, I spoke to Nixon, who had lived here for twenty years. Raised in a poor family, he was diagnosed with leprosy in his late teens. Keeping the secret came at the cost of avoiding treatment; his facial features began to thicken and he started to lose toes. The condition was physically painless but, as he remarked, 'I would rather have felt everything than feel the agony of the shame.' His father brought him hundreds of miles to the centre. After a few sporadic visits from his family, Nixon never saw them again.

In 1873 the Norwegian physician G. H. Armauer Hansen finally identified the causative agent of leprosy.[13] One of history's most notorious and divisive diseases was found to be a chronic bacterial infection of *Mycobacterium leprae*, which presented with hypopigmented (usually white) patches on the skin accompanied by sensory nerve damage. *Mycobacterium leprae* is a curious character, a bacterium that is both delicate and notoriously deceptive, hiding from the immune system by living inside Schwann cells (the insulation of our nervous wiring) or even inside macrophages, our own immune cells. After moving into a human host in search of a home, these pathogens are very pernickety about postcodes. They prefer the cooler climes of human peripheries, so they populate the nerves in the skin. In fact, their penchant for the cold means that one of the only other known reservoirs of *M. leprae* is the nine-banded armadillo, as this little armoured animal shares the same cool body temperature as human skin.[14] *Mycobacterium leprae* are also notoriously slow growers, their population taking about fourteen days to double in size, compared to thirty minutes for *S. aureus* on the skin or eighteen minutes for *E. coli* in the gut.

This fussy and leisurely bacterium is one of the few we are completely unable to grow in a laboratory, so despite our current ability to completely cure leprosy with antibiotics, this ancient disease is still shrouded in myth and mystery. Contrary to popular belief, the disease does not cause fingers and limbs to drop off. Instead, an infected human first begins to lose their sense of temperature, then the sensation of light touch, and then pain. Without the alarm of agony, sufferers damage their outer covering in cuts and burns, the resulting infections often causing permanent damage to fingers, toes and facial structures. Another commonly held myth is that leprosy is highly infectious, whereas it is actually one of the least transmissible infectious diseases and 95 per cent of people are naturally immune to it.[15] Perhaps it is the fact that this disease has the unusual combination of being both permanently deforming and contagious (to an extent) that explains why it has terrified humankind for millennia.

Leviticus 13:46 states that the leper 'is unclean, and he shall dwell alone; his dwelling shall be outside the camp'. The Hebrew word for leprosy, *tsara*, in many cases did not actually describe leprosy as we know it, but other visible skin diseases such as psoriasis and vitiligo. The common denominator was that these diseases were not simply damaging, but 'defiling', and their bearers were deemed unclean and impure. *Tsara* can also be loosely translated as being humiliated or struck down by God. Physical imperfection of the skin was a symptom of humankind's rejection of God, so lepers would have been isolated from citizens as much for spiritual quarantine as for physical. The impure were to be kept away from their camp for at least seven days and routinely checked by a priest to ascertain whether they were 'clean' and able to return. In ancient Hindu texts, including the *Atharva Veda* and *Laws of Manu*, lepers were to be quarantined and the presence of the disease was seen as punishment for past personal or family sins.[16] It was no coincidence that the first person Jesus healed was a leper.

In AD 379 the Archbishop of Constantinople, St Gregory of Nazianzus, declared lepers 'men already dead except for sin'. Medieval times were not much better for those afflicted with this disfiguring disease. While the common belief that lepers in medieval Europe were cut off from society in leper hospitals is not entirely true, stigma and strange beliefs abounded. Because of their skin, lepers were considered to be the living dead, roaming the earth with no legal rights over land or property. Understandably, many sufferers would retreat from towns to leprosy hospitals, and those who stayed would carry a small bell with them, clanging to warn the townspeople of their arrival. In some cases lepers were strangely revered, as some believed they were humans in purgatory, paying for the sins of their past life and reaching salvation through suffering; meanwhile physicians and priests who dared to touch and treat these people were sometimes venerated as saints.

The Order of Saint Lazarus was set up in AD 1119 by crusaders in Jerusalem who had set up a hospital for lepers, the order's name reflecting the Biblical man raised from the dead by Jesus.

One would think that Armauer Hansen's discovery, that leprosy was a bacterial infection, would have reduced the physical, social and spiritual stigma associated with the disease, but in fact it had the opposite effect. European colonizers and travellers found it to be endemic in impoverished people – people often assumed to be poor through moral degeneration – imbuing an already stigmatized disease with connotations of moral and sexual deviance that supposedly spread with the bacteria. The title of missionary H. P. Wright's 1889 book *Leprosy, An Imperial Danger* encapsulates the fear of moral and physical contagion.[17] In British India, lepers were isolated from communities on a systematic, institutionalized scale, and these leper colonies – or leprosaria – sprang up across the world. One of these was on the island of Molokai in the Kingdom of Hawaii and was built after the local population became ravaged by this new disease carried by Chinese and Europeans. In 1873 a Roman Catholic priest and missionary called Father Damien arrived at the leprosarium from Belgium. While almost no other uninfected European would have contemplated the idea, Father Damien decided to live with the lepers, dressing their wounds and eating from the same bowls.[18] He eventually contracted leprosy himself and died in 1889 at the age of forty-nine. As well as being recognized by the Catholic Church as a saint in 2009, he left a legacy of sacrifice and care for the poor and suffering that gave birth to thousands of charities.

Leprosy is by all measures and observable means a physical condition, but throughout history it has probably been humanity's archetypal social disease. It is remarkable that it is still commonly thought to be highly infectious yet is in fact one of the least contagious of all transmissible diseases. Even today this fact is little known and humankind continues to go to remarkable lengths to make sure that lepers live 'outside the camp'. When I reflected on my long and convoluted journey to the leprosy centre, I recalled the blank faces of locals as I asked for directions. The locals would rather not know. In every age, and wherever it has been found, leprosy has been associated with moral deficiency. The skin's ability to define the individual even permeates language, with the widespread (and often derogatory) use of the word 'leper' showing that these patients have been, and sometimes still are, defined by their disease. 'Hasn't leprosy essentially been eradicated?' is the refrain I hear when I bring up the disease with doctors in Britain. It is true that leprosy can be cured with a six- or twelve-month antibiotic regime, and effective introductions of these treatments in developing countries in the past few decades have seen numbers plummet. Nonetheless it is estimated that there are currently more than 200,000 people with leprosy around the world, and this figure may wildly underestimate the real numbers. This is because people still refuse to come forward, due to the enormous stigma that still remains in many societies – stigma that would not exist if leprosy were not a visible skin disease.

The attempt to cover up what society deems to be a curse or transgression is common, then; but the dividing power of the skin has also driven many to want to change their own perfectly healthy outer surface. I travelled east from the leprosy hospital to the sprawling East African port of Dar es Salaam, the largest city in the Swahili-speaking world. As far as the eye can see, new apartment blocks, many abandoned mid-construction, sprout up above the shantytowns like a nascent maize field. It was a good opportunity to explore a growing phenomenon across the world: skin bleaching. Skin whitening has been practised for centuries in Europe to enable the wealthy to distance themselves from field and farm. It was first mentioned by William Horman, a prominent fifteenth-century English schoolmaster, who noted that women would 'whyte theyr face with cerusse [white lead and vinegar]' for that very purpose.[19] This medieval procedure has now migrated to the modern developing world.

One in three women (and a growing number of men) in the cities of Sub-Saharan Africa regularly apply caustic creams designed to lighten black skin. Many of these are banned, resulting in the widespread use of even more hazardous counterfeits, including mercury-containing products that can cause renal failure and psychosis. In 2019 Rwandan police raided beauticians and chemists across the country in an effort to stem the thriving illegal trade of skin lightening products.[20] But the effects of the creams are not just physical. The column inches and radio minutes given to skin bleaching in Africa indicate that it constitutes a societal crisis.

I met with Camille, a local student, who revealed that more than half her friends bleach their skin. 'Why are young people ashamed to be black?' she asked rhetorically. 'You see it all the time – models on billboards and light-skinned singers in *bongo flava* (Tanzanian hip hop) videos . . . To be beautiful now is to be light-skinned.'

Camille told me that, in her circle of friends, bleaching was not so much to do with trying to be more 'Western', but rather an attempt to shake off a past of poverty. Many rural newcomers to Dar es Salaam lighten their skin to hide the shame of years in the fields, exposed to the sun. A local doctor despondently observed: 'In Africa it is not just the chameleons that change colour to survive.'

The skin is where the individual meets the group, where biology rubs up against culture. While our layer of hide is a defensive barrier against all sorts of threats, skin's social power has too often made it a weapon. Looking through the prism of skin, the darker aspects of human nature are easily observable, and to some extent we all have an ability to create the 'other'. But the wonderful paradox of skin is that the more we discover about the science and beauty of our most human organ, the more we see that we really are no better or worse underneath.

10

Spiritual Skin

How skin shapes the way we think: religion, philosophy and language

'[Skin is] substance, vehicle and metaphor'

PROFESSOR STEVEN CONNOR[1]

I FIRST REALIZED THE power of skin in religion on a visit to Kolkata, India. There is nowhere on earth like this sprawling metropolis of contrasts, which is simultaneously booming and broken. The cultural and commercial heart of eastern India is also the home of desperate poverty, where the palatial penthouses of India's new millionaires look out over squalid slums. One evening I wandered away from the monumental Victorian buildings of the city centre in an attempt to explore the 'real' city. After emerging from a side street on to a large, congested road, a fleeting scene caught my attention. On the dusty roadside across the street, a completely naked Hindu *sadhu* (holy man) was sitting cross-legged, lost in meditation, despite the hustle and bustle of the city. For just the briefest of moments, my view of the man was eclipsed by a gaggle of Muslim women wandering by, anonymous under the second skin of their black burqas. Both were demonstrating their faith: one with all skin on show, the others with none. Even the simple matter of how much skin we reveal to others is deeply laden with spirituality. We are all born naked, our skin exposed to the world, but most of us are born into culturally or religiously prescribed circumstances that mean we then spend most of our lives with some form of covering. Whether it be organized religion, informal spirituality or individual moral values, deeply held beliefs have a direct impact on the skin, and to some extent are ruled by it. Other, perhaps, than the brain, humans imbue no organ with as great a sense of holiness as the skin. It has preoccupied theologians and fascinated philosophers, and influences our own everyday thinking in unexpected ways.

Among the human organs, skin has a special place in religion as it uniquely combines two characteristics critical for spirituality: physical space and sensation. For Navajo Native Americans, their skin literally defines their place in the world. The ridges in the skin of their fingers and toes literally anchor them between earth and heavens: 'These whorls at the tips of our toes hold us to the earth. Those at our fingertips hold us to the sky. Because of these we do not fall when we move about.'[2] Indeed, for all of us, our barrier organ divides our inner self from everything else in the universe. It is simultaneously a barrier *from* and the contact point *with* the rest of the world. Dr Thomas A. Tweed, Professor of Religion at the University of Notre Dame, argues that 'religions are about dwelling and crossing, about finding a place and moving across space'.[3] We see this in pilgrimages to specific geographical locations, the distinct layout of religious buildings and a follower's movement from this world to the afterlife. Skin is akin to a curtain veiling a sacred space, a wall enclosing our inner bodily temple. Breaking through this wall is a deeply profound undertaking. Within the span of a single week I found myself talking to two different people who had cut into human skin. One was a liver surgeon, while the other was a gang member in the emergency department, recently arrested for knife crime. Both vividly recounted the first time they cut into skin, and even though it was for utterly different reasons, both felt as if they had crossed a sacred boundary and moved into a forbidden space.

The skin acts as a barrier between flesh and the world, yet it is also the critical part of us that yields to the desires of the flesh. The skin is a sensory organ, indeed our largest sexual organ, infused with a heady mixture of desire, sin and shame. Skin is associated with the human fragilities of sexuality and mortality; its nakedness is tied to the corruption of the human soul, as we see from the story of Adam and Eve in Genesis. Before the Fall they are shamelessly naked, and after the Fall they feel the need to cover themselves.

Self-inflicted pain and skin mutilation often represent religious self-denial and the putting-to-death of the physical flesh, whether this be self-flagellating Catholic monks or Hindu devotees dragging chariots attached to hooks lodged in the skin of their backs during the Thaipusam festival. The pain associated with the etching of tribal tattoos has a religious dimension, encapsulating the religious journey through struggle to triumph. In the Bible, the worst physical punishment that Satan could conceive for Job was an insufferable itch; the urge brought about by the itch, and the intense but short-lived pleasure of scratching, is a metaphor for temptation. The Qur'an, Surah 4:56, recognizes the exquisitely sensitive thermoreceptors in our skin and this organ's ability to bring about excruciating pain: 'Those who disbelieve in Our verses, we will drive them into a fire, every time their skins are roasted / We will replace their skins with other new skins so that they may taste the torture.'

As our most visible organ, skin is often central to religious identity. By means of the often-overlooked communication of clothing, both the Hindu *sadhu*'s naked, exposed skin and the completely covered Muslim women on the street in Kolkata were expressing piety, submission and identity. After Adam and Eve's sin, nakedness lost its innocence and became a sign of guilt and shame, and body modesty is still central to the orthodox teachings of the Abrahamic religions.[4] Given that God actively clothed Adam and Eve with animal skins after their sin, Christian theologian John Piper suggests that coverings have both a negative and positive spiritual purpose: 'God ordains clothes to witness the glory we have lost . . . but it is also a testimony that one day God would make us what we should be'.[5] This need for clothes starkly differs from the neo-pagan ideals of being 'skyclad', a state of regular ritual nakedness where one is clothed only by the sky; removing clothes removes barriers between themselves and their gods above and the Earth below.

Temporary body art has been used in belief as well as beautification since time immemorial. Hindu women on the Indian subcontinent have long worn *bindi*, a red spot between the eyebrows marking a sacred point, the 'third eye', which represents a higher, invisible, state of consciousness. Permanent tattoos, from the crude keloid scars that give crocodile worshippers of Papua New Guinea reptilian skin, to the intricate and protective *yantra* tattoos on the backs of Southeast Asian Buddhists, have long been used in religion. Religious tattoos not only seek to show identity, but also strengthen the skin's protection against spirits and make the body recognizable for the afterlife. The Lakota people of the Great Plains of North America, for instance, have traditionally believed that they need individual tattoos marked on their skin so that when they die they can be recognized by an old woman, called the 'Owl Maker', who will let them into the plentiful hunting grounds of the afterlife.[6]

Religious and cultural ceremonies are bound by time and briefly reflect milestones of faith and life, so altering the skin need not always be permanent. Perhaps the most famous markings to temporarily transform one's body are the intricate reddish-brown markings of mehndi, commonly called henna. This paste, made from the dried and ground leaves of the henna tree, dyes the top layers of the skin until the entire epidermis has been replaced, a few weeks later. The plant probably originated in Ancient Egypt and travelled to India, where the practice of body art, particularly to adorn brides for marriage, has been used for rituals and celebrations for millennia. Indeed, its widespread use is seen in the earliest Hindu scriptures.

In some religions, however, skin markings are most notable in their absence. The majority of Islamic scholars believe tattoos to be *haram* (forbidden) as they mutilate the body and change God's creation. The Jewish scriptures prohibit cutting and tattooing, and although most Christian denominations do not see these laws as binding, missionaries and popes alike have discouraged the marking of the skin for centuries. One form of skin modification, however, is utterly essential to Jewish identity. Male circumcision, involving the removing of the foreskin, is carried out on the eighth day of life and is a physical sign of a covenant between God and Abraham's descendants. It became a matter of contention for the emerging Christian church in the New Testament, the apostle Paul, for instance, arguing that there is no longer a need for physical circumcision because Christian believers undergo a spiritual 'circumcision of the heart'. The progression is from the physical to the transcendent, meaning that the old rite is no longer necessary.

In modern developed society, where most people have access to showers and shampoo, the notion of dirty, unclean skin has largely been lost. It is not surprising, though, that dirty skin has been historically associated not just with poverty, but also with the spiritual corruption of the inner person. At the heart of many religions, where impure skin represents an impure soul, lies ritual purification. When I visited the great mosques of Cairo and Istanbul, I was struck by the centrality and beauty of the washing fountains in the main courtyards. The Prophet Mohammed himself said, 'cleanliness is half of faith'. *Wudu*, the ablutions of hands, feet and face before Muslims' Friday prayers, entails the physical washing of the external skin to represent the purification of the heart. Adherents of the Japanese Shinto faith purify themselves by carrying out *misogi*: plunging their naked bodies underneath waterfalls or into the sea. Baptism with water – something I have experienced as a Christian – is a physical sign of washing that represents the core of the Christian faith: dying to an old life of wrongdoing and being born again in the new life of Jesus Christ.

The ritual purification of the skin concerns not only physical cleansing from dirt, but also separation from illness, corruption and disease. A number of years ago one of my friends developed the scaly plaques of psoriasis on her elbows and across her abdomen. She tried every cream and medication on the market to overcome a disease that was shattering her self-confidence. She burst into tears when she told me she hadn't been to a beach or swimming pool in three years. I encouraged her to see a dermatologist; in recent years, revolutionary 'biologic' medications that target specific molecules in the immune system have transformed psoriasis outcomes, and I was convinced that she would benefit from them. When I caught up with her six months later, she had been completely cured. I couldn't wait to hear which of my favourite new treatments had done the job. But instead of seeing a dermatologist, she had visited a druid. After a bewildering foray into her discovery of neo-paganism, she described how a few sessions of hypnosis and meditation had miraculously cured her of her skin disease. Skin is a physical link between mind and matter, and it is certainly true that, in some cases, stress relief and the altered states of consciousness associated with meditation and spiritual experience appear to have an effect in 'purifying' the skin. I would not recommend meditation as a first-line treatment for psoriasis, but the mystery between our mind and skin only strengthens the skin's position as a transcendent organ.

Perhaps the most profound window into skin's religious power can be seen on the altar wall of the Sistine Chapel, in Michelangelo's *The Last Judgement*. Near the centre of this enormous fresco, looking up hopefully at Jesus, is St Bartholomew. In one hand he holds the knife that skinned and martyred him, and in the other dangles the limp, empty cloak of his flayed skin. A closer look unveils a cryptic optical illusion, where the flabby folds of St Bartholomew's skin make up arguably the only self-portrait of Michelangelo in all of his art. But why did this master artist project himself on to this ugly envelope of flesh? In the uncertain landscape of *The Last Judgement*,

Michelangelo hopes that Christ will have mercy on him; St Bartholomew is therefore offering the only part of the artist's body that could identify him as he seeks a new body in heaven. It's the same belief as that of the Naga 'tiger warriors', who see their tattoos as their only possessions, as these are the only addition to their sense of self that they can carry on to the afterlife. A body rent of its skin, like the écorché statues at the beginning of the book, resembles a human, but not a person. Skin is synonymous with soul. Religion shows that, even when lifeless and separated from the rest of our body, our skin is the essence of us.

Even to the non-religious, skin is deeply philosophical. It has the sense of the divine, and we have all experienced the numinous on our surface; blushes of embarrassment, indescribable sexual touch and involuntary shivers while listening to a powerful piece of music are just a few ways in which skin takes us to a higher place. As skin is so intertwined with our very being, and

THE LAST JUDGEMENT

mediates our relationship with the outside and inside, it has taken on forms very different to its physical appearance.[7] Humans have long mused over the supernatural meanings of skin, and to get a brief grasp of this critical conversation, we can call in the help of three French philosophers.

Didier Anzieu, one of the great psychoanalysts, dedicated much of his life to his concept of the 'skin-ego'. He argued that the body's surface is an integral part of the functions of the mind. Anzieu attempted to put into words the symbolic skin we imagine to surround us. In the same way that our physical skin envelops our body, we all have an idea of our skin as some form of psychic envelope that wraps around our mental make-up. Building on Sigmund Freud's concept of the ego, Anzieu describes the skin-ego as 'a mental image of which the Ego of the child makes use during the early phases of development to represent itself as an Ego containing psychical contents, on the basis of its experience of the surface of the body.'[8] Anzieu's abstract concept reflects the functions of our physical skin: the skin-ego contains our thoughts and feelings, protects us from other ideas and egos, communicates with the world, initiates sexual feeling and separates us as individuals. A baby has almost no knowledge of where its body, bound by skin, ends and another person begins; indeed, it often feels as though it shares the skin of its mother. As the child develops, it builds a concept of being contained by its own skin and thus of its own personhood and individuality. Now that the infant has acquired a 'skin-ego' it is able to translate physical sensations from the skin to make sense of its own psychological framework. Is a touch hurtful or loving? Being able to interpret this marks the point at which a child has both a physical and mental skin. Alongside the idea of containment, Anzieu believed that the skin-ego provided two additional functions, those of protection and inscription. In the first instance the skin defines us as distinctive individuals, and in the second it communicates that individuality to others.

Abstract rather than scientific, the skin-ego is a particularly compelling concept in relation to the spectrum of personality disorders. In those who have a narcissistic personality disorder, their psychic envelope can be seen as being pathologically thickened. Not only does this double wall of thickened skin give the narcissist a feeling of invincibility and superiority but it also reduces their ability to empathetically 'feel' others. A narcissist is often described as 'callous', a direct reference to the thickening of the epidermis. At the other end of the spectrum is emotionally unstable personality disorder, also known as borderline personality disorder. With a disturbed sense of identity, fear of abandonment and unstable emotional responses, the skin-ego is weak, broken and porous. Anzieu likened 'borderline personality' skin to 'an egg with a broken shell emptying itself of its white'. The damage to this imaginary surface can be reflected on the physical skin; borderline personalities often predispose to self-harm and mutilation.

The way in which skin separates space has also influenced modern philosophy. Skin is the house in which our physical and mental selves are contained, so it acts both as a wall to block off the outside world and as a window to let it in. This dual role is put beautifully by Gaston Bachelard in his seminal book *The Poetics of Space*:

> '. . . on the surface of being, in that region where being wants to be both visible and hidden, the movements of opening and closing are so numerous, so frequently inverted, and so charged with hesitation, that we could conclude on the following formula: man is a half-open being.'[9]

Our third and final French philosopher, Michel Foucault, took the idea of philosophical skin further, to see how societal power influences the human body and the concept of identity. He came to recognize that on both an individual and societal level, our physical skin is intertwined with our very being. Foucault argued that any intentional physical change to the appearance of the skin, from Botox to body art, is a 'technology of the self'.[10] We change our bodies 'in order to attain a certain state of happiness, purity, perfection or immortality'. When we change our skin, we change ourselves.

These philosophical ideas make it hard to deny that as well as being physical, skin is also imaginary and fantastical. Another common way in which skin is expressed as metaphor is in its comparison to a book, a chronicle containing the story of our lives. We all have some sense that our skin is a parchment upon which colours, scars and wrinkles tell our history. But it is not all written in indelible ink; the skin also communicates as a palimpsest, a wax tablet on which we, the scribe, scrape away the old writings to reuse again and again. Our superficial story is partly biography, reflecting ancestry and age, revealing health and disease and spilling secrets through blushing and sweating. As humans, this hasn't stopped us trying to change the narrative, of course. We adjust our skin colour, from the relatively recently perceived ideal of the 'healthy tan' in the West to the exponential rise of skin bleaching in other parts of the world. The idea of skin 'giving away your age' reveals that skin holds secrets and when we hear the phrase 'anti-ageing', the first thing we think of is our skin. Humans also try to write their skin as autobiography by covering it, painting it and permanently drawing on it: the most intimate way of showing who we are, and who we want to be. If skin speaks about our past and our present, it's no surprise that palm-reading – inferring someone's future through the skin creases of their hand – has long been a ubiquitously popular (albeit completely unscientific) way of trying to look into one's future. But just as any form of communication can become a means for persecution, it's clear that humans have also tried to control the skin story of others.

With surprising frequency throughout history, the 'book' concept of skin has even surpassed metaphor and become grisly reality. When I visited the museums at the Surgeons' Hall, the grand headquarters of the Royal College of Surgeons of Edinburgh, an elegant pocket book in one of the displays caught my eye. I leaned closer to look at the dark-brown leather, assuming it to be the notes of a celebrated surgeon. Instead I found the faded words imprinted on the cover: 'Burke's Skin Pocket Book'.

In 1828, Edinburgh was running out of dead bodies. With the advent of modern surgery, the city had become the world capital for the teaching of anatomy, led by Professor Alexander Monro and the popular Scottish anatomist Dr Robert Knox. After national crackdowns on grave-robbing, demand for bodies from which budding surgeons could learn their trade vastly outstripped supply. When one of William Hare's lodgers died of dropsy, he planned to make up for the loss of rent by taking the body to Knox, with the help of his friend William Burke. When the doctor agreed to take the corpse for a whopping £7 (more than £700 in today's money), and one of Knox's assistants said he 'would be glad to see them again when they had another to dispose of', Burke and Hare knew they were on to something.[11] Over the coming months, the pair murdered sixteen victims, providing Dr Knox with a fresh supply of bodies. Burke was eventually caught in 1829 and hanged in front of 25,000 spectators. Fittingly, Professor Monro dissected his body in front of a packed anatomy theatre in Edinburgh Medical School. In a process called anthropodermic bibliopegy, Burke's skin was stripped, tanned and now binds the pocket book that can be seen today.[12] Seventeen other books around the world are known to be bound in human skin, although many more await testing. At the border of the body, skin is simultaneously inside and outside. Our concept of it as a book not only manifests as the pages of our life story, but also as the cover.

Finally, one of the most common ways in which our metaphorical skin transcends our physical skin is in everyday language. Being 'thick-skinned' and having someone 'get under your skin' play on the many themes of our skin's boundary metaphors. 'I'm touched' and 'you've hurt my feelings' derive from the emotional power of our sensing skin, with 'callous' and 'tactless' as words to describe someone unable to metaphorically 'feel' others. When it comes to the temporary changing of our skin story through 'make-up', the word itself is incredibly revealing: in changing the appearance of our outer layer we can be changing – indeed making – our very self. But the core phrases relating to skin, common to most languages, reveal something unique, and very human, about this organ. On the one hand, some idioms of the integument, such as 'skin-deep', suggest that the skin is superficial and trivial. But in most other phrases skin is placed at the centre of our very being: 'saving one's skin'; being 'comfortable in your own skin'; getting 'into', 'under' or 'jumping out of' one's skin. The French *vouloir la peau* ('to take someone's life') and the Italian *salvare la pelle* ('to save one's skin') indicate that equating our skin with our very being is universal; indeed, in these phrases the word 'skin' is a substitute for one's self. This paradox of skin being simultaneously nothing and everything mirrors the tension in the relationship we have with our outside organ, as well as with our own human condition.

Our skin is not only a physical presence; it is an idea. In the same way that our physical skin contains us while we try to contain it, what it represents has directed the course of history and profoundly affects our own life. Our forgotten organ has long been seen as wrapping paper, stripped away to make the écorché statues of 'serious' medicine. But the more we look, the more we see that what is found at our periphery is actually the centre of what it is to be human. Our skin is our self.

Glossary

Acne

Officially called acne vulgaris (*vulgaris* being the Latin for 'common'), acne is a skin condition characterized by various types of spots (papules, pustules and nodules) and inflamed skin. It is caused by a complex cocktail of genetics, hormones and environmental factors. As this highly visible condition usually erupts during formative teenage years, it also has vastly underestimated psychological and social implications.

Adermatoglyphia

A vanishingly rare genetic disorder – only observed in five families across the world – resulting in the complete absence of fingerprints. After a Swiss woman was denied entry to the US until a specialist dermatologist could prove her condition, it has also been termed 'Immigration Delay Disease'.

Adipocyte

Fat-containing cells found in great abundance just beneath the dermis. These are indispensible reservoirs of energy for the human body.

AGEs (advance glycation end products)

Proteins and lipids in the body that have been modified by the bonding of sugar molecules. The acronym is particularly fitting, as AGEs are associated with age-related diseases, including type 2 diabetes and heart disease.

Allodynia

The sensitization of an area of body tissue, lowering its pain threshold. Often caused by damage or inflammation. Think of putting a shirt on to a sunburned back.

Alpha hydroxy acids

A group of chemicals (including lactic and citric acids) commonly used in skin peels. They reduce cell adhesion in the outer layer of the epidermis, stimulating exfoliation.

Anthropodermic bibliopegy

A dignified name for a grisly process: binding books in human skin.

Antioxidant

Molecules that inhibit a chemical reaction called oxidation, which produces chemically reactive and tissue-damaging molecules called free radicals. Their purported role in preventing disease is intensely disputed within the scientific community.

Apocrine gland

Sweat glands located in the armpits, groin and nipples, secreting oils rich in proteins, fats and pheromones. Unlike eccrine glands, they produce sweat in short bursts, stimulated by adrenaline. These are the producers of what could be called emotional sweat, from fear to sexual arousal.

Archaea

A little known but ubiquitous group of microbes that physically resemble bacteria, but are genetically completely distinct. They contribute to the nitrogen and carbon cycles of both the Earth and the human body, and none are known to cause human disease.

Arrector pili muscle

Tiny muscles attached to hair follicles that, when contacted, cause hairs to stand on end.

Atopic dermatitis (*see* eczema)

Autonomic nervous system

A part of the human nervous system that unconsciously influences our internal organs, from gut movements to the stimulation of the fight-or-flight response.

B cell
The immune cell that is responsible for producing antibodies against foreign molecules. They reside in lymph nodes, and are able to engulf pathogens that invade the body, and present their epitope (or bacterial barcode) on their surface. If this epitope on the surface of the B cell is recognized by a T cell in the lymph node, the T cell provides a signal to the B cell. This enables it to transition into a 'plasma cell', which is essentially an antibody factory, producing antibodies targeted at the pathogen.

Bacillus oleronius
Bacteria that reside inside mites and termites, including the Demodex mite. When this mite dies on human skin it releases *Bacillus oleronius* into the skin, which is thought to cause an immune reaction that manifests as acne rosacea.

Basal cell carcinoma
The most common, and one of the least dangerous, forms of skin cancer. It classically presents as a shiny bump on sun-exposed areas of skin.

β-endorphin
A molecule produced within the body that binds to the same receptor as opium. Plays a crucial role in pleasure, reward behaviour and addiction.

Bilirubin
A yellow molecule produced by the breakdown of red blood cells. It is most well known for causing the yellow discolouration of jaundice on the skin, but it can be more commonly seen when bruises turn yellow after a few days.

Brain natriuretic peptide
A hormone that, despite its name, exerts most of its actions on blood vessels. Its most studied function is to lower blood pressure by dilating peripheral blood vessels.

Carotenoids
Pigments found in plants, algae and bacteria that create red, orange and yellow colours. They convey a diverse range of health benefits, and colourful fruit and vegetables are vital for a healthy, balanced diet.

Catechin
Chemicals found in plants, most notably green tea and cocoa, that have been shown to have antioxidant, anti-inflammatory and anti-cancer properties in the lab, but the evidence for their effect on preventing human disease is mixed.

Cerebellum
The lower area of the brain, critical for the motor functions of the body, including voluntary movement, balance and coordination.

***Clostridium difficile* infection**
An infection of the gastrointestinal tract that presents with abdominal pain and watery diarrhoea, and can result in severely dilated or perforated intestines and life-threatening sepsis. It is a hospital-acquired infection, spread by bacterial spores of the bacterium *Clostridium difficile* in faeces, and has been a driving force behind campaigns for thorough hand-washing, cleaning and antibiotic stewardship in health-care environments.

Coeliac disease
An autoimmune disease in which the body's immune system reacts to gluten and damages the mucosal lining of the intestines, causing mal-absorption and diarrhoea. A gluten-free diet is the only current treatment.

Collagen
The most abundant protein in the human body, which acts as a structural scaffold for most tissues. Type 1 collagen, the most common form of this protein, forms large, rope-like fibres in the dermis that give human skin its structure.

Commensalism
A biological relationship where one organism gains benefit and the other neither benefits nor is harmed.

Congenital insensitivity to pain
A rare genetic condition where sufferers cannot perceive physical pain. Sensation is still intact, so they can perceive if something is rough or smooth, hot or cold. One cause is a mutation disabling sodium channels in pain-detecting nerves, so that pain signals from the periphery never reach the brain.

Crusted (Norwegian) scabies
A severe form of scabies, where scabies mites have been able to thrive and multiply on the skin of those with weakened immune systems, such as the elderly. Sufferers can become reservoirs of thousands of mites, and are extremely infectious.

Cytokine
A small protein that acts as a messenger between cells of the body, and is particularly crucial in the human immune system. This is abundantly clear in the remarkable new treatments that target inflammatory cytokines in autoimmune diseases, revolutionizing the treatment of psoriasis, Crohn's disease, rheumatoid arthritis and multiple sclerosis.

Dermatome
A discrete area of skin, the sensation of which is supplied by one nerve from the spine. There are thirty dermatomes from top to toe.

Dermis
The layer of skin below the epidermis and above the hypodermis (or subcutaneous tissue). It plays innumerable roles in the functions of the skin and the body. See Chapter 1 to get a glimpse of its glories.

Eccrine gland
The most common form of sweat gland, found on all surfaces of skin. They secrete sweat in response to raised body temperature, apart from those on the palms and soles, which respond to emotional arousal.

Écorché
A depiction of the human body without its skin.

Eczema
A commonly used word for atopic dermatitis, a chronic itchy skin condition. It has various complex causes, which usually boil down to skin barrier dysfunction and immune dysregulation. Both of these influence each other in a spiral of itching, scratching and misery.

Elastin
As its name suggests, this is a stretchy protein that is responsible for giving skin its shape after being pressed or stretched.

Enterotoxin B
A powerful toxin produced by *Staph aureus* bacteria that causes inflammatory responses in the body, leading to dermatitis, food poisoning and sometimes deadly toxic shock syndrome.

Epidermis
The outermost layer of skin, responsible for most of the skin's barrier functions.

Epidermolysis bullosa
A collection of genetic disorders that cause easy blistering of the skin. There is currently no known cure, although a successful genetically altered skin-graft treatment in 2017 suggests this may be about to change.

Epigenetics
The study of changes in gene expression that do not alter the genetic code itself. In essence, how genes are turned on or off.

Epitope
A part of an antigen (any structure that can be bound by an antibody) that is recognized by the immune system. Think of this as the specific barcode of any microbe that enters the body, and how the immune system recognizes foreign pathogens.

Exfoliatin
Another *Staph aureus*-derived toxin that breaks down specific adhesion proteins in the skin, causing skin to break apart and let bacteria in.

Extracellular matrix
A diverse meshwork of molecules that links cells in the body to each other, both structurally and biochemically.

Extrinsic ageing
Ageing of the skin caused by external factors, including sunlight, diet, smoking and air pollution.

Fibroblast
Cells found in the dermis that produce the vital structural proteins collagen and elastin, as well as other molecules critical for the functioning of the extracellular matrix.

Fibrosis
The production of excess connective tissue. When this is in response to injury, it is known as scarring.

Filaggrin
A protein crucial for the healthy barrier functions of the epidermis. Mutations in the gene encoding filaggin have recently been found to be responsible for at least half of eczema cases.

Furanocoumarins
Molecules naturally produced by certain plants, including wild celery and cow parsley, which damage DNA in skin cells when exposed to ultraviolet light. Skin contact with furanocoumarins, followed by exposure to sunlight, brings about intense inflammation and blistering. This is likely to be the plant's defence mechanism against hungry animals.

Glabrous skin
Hairless skin. Most commonly refers to the palm of the hand and sole of the foot.

Glutamine

An amino acid that is an essential building block of many proteins, and has legion other uses, from the generation of cellular energy to regulating the body's nitrogen and ammonia cycles.

Glycaemic index

A system that ranks carbohydrate-containing foods by how quickly they affect blood sugar levels. For example, sugary drinks and white bread have a high glycaemic index, whereas most vegetables and grains have a low glycaemic index. Cooking and food processing often increase the glycaemic index of food.

Glycosaminoglycans

The ground substance of the extracellular matrix, giving it structure while allowing cells and molecules to move around. But it's also so much more than a ground substance, playing active roles in skin healing, inflammation and wound repair.

Harlequin ichthyosis

A rare, life-threatening genetic disorder resulting in hard, cracked skin. This gives us a sobering insight into how the skin's barrier function is essential for human survival.

HEV light

High-energy visible light. The highest energy wavelengths of light within the visible spectrum, namely blue and violet.

HIFs

Hypoxia-inducible factors. Proteins that alter the rate of DNA expression in response to low levels of oxygen.

Histamine

A very small compound that packs a big punch. When released by mast cells it causes many of the symptoms of inflammation and allergy: itching, blood-vessel dilation leading to redness, heat and swelling on the skin, and sometimes a systemic drop in blood pressure. It also causes sneezing and increases nasal secretions.

Hygiene hypothesis
The well-supported theory that the increased environmental cleanliness in the modern world reduces childhood exposure to microorganisms and infectious disease, stunting the normal development of immune systems. This could be an explanation for increasing levels of allergy across the world, particularly in developed areas.

Hyperkeratosis
An excessive formation of keratin in the outer layers of the epidermis.

Hypodermis
Also known as subcutaneous tissue, the layer immediately below the dermis, largely consisting of adipocytes and fibrous bands of collagen. It is not usually considered to be a layer of the skin.

Hypothalamus
An almond-sized part of the brain that is, among its many other complex functions, the primary link between the brain and the body's hormonal system. With relevance to the skin, it is the body's thermostat, its central circadian clock and a key bridge between the mind's formulation of fear and stress and its physical effects on the body.

Immune tolerance
A mechanism by which the immune system does not react to specific tissues or substances. It is very important that the immune system does not attack 'self' tissues, and a failure in immune tolerance often results in autoimmune disease.

Innate lymphoid cell
A recently discovered family of immune cells that plays a role in both quickly responding to infectious organisms and regulating the body's frontline immune response in the skin, gut and airways.

Intrinsic ageing
Also called chronological ageing. This is the natural course of skin-ageing over time, most notably the steady depletion of collagen in the dermis, from roughly the age of twenty onwards.

In vitro
Literally 'within the glass'. Scientific experiments carried out in the laboratory, often in test tubes and dishes, outside a living organism.

In vivo
Literally 'within the living'. Scientific experiments carried out using whole, living organisms.

Kangaroo care
Skin-to-skin contact between a newborn baby and its mother (or other caregiver).

Keratin
A tough, fibrous protein that makes up the outer layer of human skin, as well as hair, nails, claws and horns.

Keratinocyte
The primary cell of the epidermis, which produces the protein keratin.

Langerhans cell
Immune cells within the epidermis that take up and process parts of microbes and display them to the effector cells of the immune system.

Leishmaniasis
A disease caused by the *Leishmania* parasite and spread by the bite of sandflies. Cutaneous leishmaniasis is characterized by wide, shallow ulcers of the skin.

Macrophage
From the Ancient Greek for 'big eater'. These immune cells are found all over the body and engulf and digest microbes or parts of microbial debris. They can then communicate information about these microbes to other immune cells, as well as directly destroying the invaders.

Major histocompatibility complex
A group of proteins on the cell surface used to present parts of foreign microbes to other immune cells. Each human has a unique set of MHC

proteins, so these are used to measure the compatibility of a piece of tissue (*histo* in Ancient Greek) from one human for another, for example in organ transplantation.

Malassezia
A type of fungus, commonly found on the skin surface of mammals.

Matrix metalloproteinases
Enzymes responsible for the degradation of proteins within the extracellular matrix.

Mechanoreceptor
Sensory receptors providing information to the brain about mechanical distortion or pressure on the skin.

Melanoma
Skin cancer that derives from melanocytes in the epidermis. It is the most dangerous type of skin cancer. Melanomas usually have a distinctive appearance, detectable using the ABCDE method (see Chapter 4), but they can also present atypically, such as the pink/red 'amelanotic melanoma'.

Microbiome
The community of trillions of microorganisms that live in and on us. Microbiomes can be found in separate surfaces or organs of the body, such as the skin microbiome or gut microbiome.

Mutualism
When two organisms of a different species interact with each other in a way that benefits both.

Myocarditis
Inflammation of the heart muscle, most often caused by viral infection, but it can also be bacterial or autoimmune. Myocarditis often presents with chest pain, palpitations and fever. The mainstay of therapy is symptomatic treatment.

Navajo people

A Native American people, today primarily located around the quadripoint of Arizona, New Mexico, Colorado and Utah.

Neuropathic pain

Pain caused by damage to nerves, resulting in abnormal excitation and pain signals being sent to the brain. Neuropathic pain can also form within the central nervous system, resulting from cellular and molecular changes in the spine or brain.

NHS

The National Health Service. The publicly funded healthcare system of the United Kingdom, although run separately by each of the four constituent countries. Founded in 1948 to be 'free at the point of use'; this is still largely the case for most services today.

Nitrogen dioxide

A compound produced by the burning of fossil fuels (in cities, most notably by motor vehicles) and cigarette smoke. It causes airway inflammation in healthy individuals and exacerbates respiratory diseases.

Nociceptor

A specialized sensory receptor that alerts us to actual or potential tissue damage, resulting in pain.

Nutrigenetics

The study of the interplay between nutrition and genetics, particularly the response of specific gene variants to food and nutrients.

Onchocerciasis (river blindness)

A disease caused by the *Onchocera volvulus* worm, characterized by insufferable itching and loss of vision. The vast majority of cases are found near Sub-Saharan African rivers, as these are the habitats of the blackfly. The blackfly bites human skin, releasing *Onchocera* larvae into the dermis and subcutaneous tissue. After maturing, the male and female worms mate and release their offspring, microfilariae, up into the skin to be taken up by blackflies dining on a blood meal. The microfilariae that do not join this next part of the life cycle die, releasing the bacterial contents of their microbiome into the skin. One of these bacteria, *Wolbachia*, causes a severe inflammatory response in human skin.

Oxytocin

A neurotransmitter molecule involved in uterine contraction during childbirth and the milk ejection reflex in breastfeeding. It also has a reputation as the love hormone, with hugging, kissing and sex stimulating its release, and subsequently influencing bonding behaviour.

Palmitoyl pentapeptides

Chemical compounds used in cosmetic research and practice (particularly palmitoyl pentapeptide-4) that can penetrate the lipid layers of the skin and stimulate the regeneration of molecules in the dermis, such as collagen.

Pathogen

An infectious agent that has the ability to cause disease in its host.

Phospholipase

An enzyme that breaks down phospholipids into fatty acids and other lipids. Recent research shows that some of these molecules can be recognized by the immune system and contribute to inflammation.

Photosensitiser

A substance that does not damage tissue on its own, but when exposed to a light source (and in the presence of oxygen) can damage chosen structures, from microbes to cancerous tissue.

Phytophotodermatitis

Inflammation of the skin caused by the interaction of UV light and plant-derived molecules.

Prostaglandin

Lipids found throughout the human body that carry out diverse functions, notably contributing to the dilation of blood vessels and inflammation.

Psoriasis
A chronic inflammatory skin condition characterized by red, dry, itchy plaques with clearly demarcated edges. Any area of skin can be involved, but it classically appears on the scalp, elbows and knees.

Postherpetic neuralgia
Pain following shingles, caused by damage to nerves by the varicella zoster virus. See 'neuropathic pain'.

Pyrolobus fumarii
A hardy species of archaea which lives in hydrothermal vents 2,000 metres below sea level at temperatures of 113°C.

Regulatory T cells
Immune cells that suppress immune responses to self-molecules, helping prevent autoimmune disease.

Relapsing fever
A disease of fever, headaches and skin rashes caused by *Borrelia recurrentis* bacteria spread by the body louse.

Ringworm
An infection of the skin, caused by a wide range of fungi and manifesting as a red, itchy, circular rash. Ringworm's medical name is 'tinea', which is usually followed by the Latin name of the body part affected, e.g. tinea capitis for the scalp and tinea pedis (athlete's foot) for the feet. It has nothing to do with worms.

Rosacea
A chronic red, pimply rash that typically affects the nose, cheeks and forehead. It is more common among Caucasians between the ages of thirty and fifty. The cause is not well understood, although contributory factors are likely to be immune dysfunction, Demodex mites, sun exposure, blood-vessel dilation and genetic factors.

Scabies
An irrepressibly itchy rash caused by the burrowing mite *Sarcoptes scabiei*. It is treated with insecticide creams applied to all areas of skin.

Schizophrenia

A chronic mental health disorder that distorts the way an individual thinks, behaves and perceives reality. Symptoms include delusions (firmly held false beliefs), hallucinations (usually hearing voices), withdrawal from social interaction and reduced emotional expression. The word 'schizophrenia' means 'splitting of the mind', which crucially is very different from split personality, which this condition is commonly misunderstood to be.

Seborrheic dermatitis

A disorder of itchy, flaky, red skin in areas highly populated with sebaceous glands, namely the face and scalp. It is caused by an overgrowth of the *Malassezia* yeast, which causes an immune response and subsequent inflammation. The seborrheic dermatitis on the head of a newborn is commonly called 'cradle cap', and the uninflamed form of seborrheic dermatitis on the scalp in adults is called dandruff.

Sebum

A slightly yellow, oily substance containing various types of fat molecules. It lubricates, acidifies and helps waterproof the skin.

Selenium

A micronutrient: essential in small quantities for human function and survival. It is commonly used in health supplements for a variety of diseases, but there is currently next to no evidence for it having any effect on reducing disease or improving mortality.

SIK inhibitor

SIK (salt inducible kinase) is a protein that regulates the production of melanin. SIK inhibitors are small molecules that block the function of this protein, increasing the production of melanin across the skin.

Spider naevus

Also known as a spider angioma, a spider naevus is a group of swollen blood vessels underneath the skin. It does not look like a spider, but rather a spider's web, with a central red spot and radiating branches. They are caused by high levels of oestrogen in the blood, itself caused by pregnancy, hormonal contraception or liver disease.

Squamous cell carcinoma
One of the three main types of skin cancer, along with basal cell carcinoma and melanoma. It often presents as a hard, scaly ulcerated lump on sun-exposed surfaces, but its appearance is highly variable. Sunlight exposure is the main risk factor, but people who are immunosuppressed (most notably those on immunosuppressive medication following a solid organ transplant) are also significantly at risk.

Staphylococcus hominis
A generally harmless bacterium (unless you count contributing to body odour harm) that is resident on the surface of the skin. However, some strains can cause infection in immunocompromised individuals.

Staphylococcal Scalded Skin Syndrome
Red, blistering skin resembling a burn, caused by the exotoxins of *Staphylococcus aureus*. The toxins damage desmosomes – the protein anchors between skin cells – so the skin begins to break apart and peel off. It mostly affects children under the age of five, as the body begins to form antibodies to the exotoxins during childhood. It can be quickly and effectively treated with antibiotics.

Synapse
A junction connecting one neuron (nerve cell) to another. Signals pass across the synapse via molecules called neurotransmitters.

T cell
A group of immune cells that are part of the adaptive immune system, in that they mount targeted responses towards specific pathogens. They are able to both directly kill cells infected with pathogens, and pass on chemical signals to other immune cells to coordinate an attack.

Trench fever
A short-lived disease of fever, headaches, skin rashes and leg pain. Caused by *Bartonella quintana* bacteria spread by the body louse.

Trimethylaminuria

A rare genetic disorder in which trimethylamine – a product of the decomposition of food inside our gut – cannot be broken down. Instead, levels increase within the body and trimethylamine is released through the affected individual's sweat and breath, giving off a powerful fishy odour.

Typhus (specifically 'epidemic typhus')

A disease of fever, headaches, skin rashes, light sensitivity and occasionally death. Caused by *Rickettsia prowazekii* bacteria spread by the body louse.

Ultraviolet (UV)

A form of electromagnetic radiation that has a shorter wavelength (and higher energy) than visible light but a longer wavelength than X-rays. Roughly 10 per cent of the radiation produced by the sun is ultraviolet.

Urushiol

Oily molecules found within certain plants – most famously poison ivy – that cause an allergic rash on human skin.

Vaginal seeding

The process of rubbing the mother's vaginal fluid across the skin of a newborn baby delivered by caesarean section. The idea is to coat newborn babies with a 'natural' community of microbes, potentially reducing the risk of disease in the future. Despite the sensible logic behind the concept, as of early 2019 there is no clear data on the long-term health consequences of vaginal seeding. It is also important to assess the risk of infecting the neonates with potentially harmful vaginal microbes such as group B streptococcus and sexually transmitted infectious agents, including *Neisseria gonorrhoeae*, *Chlamydia trachomatis* and the herpes simplex virus.

Vector

A disease vector is an agent (living or inanimate) that transmits an infectious pathogen into a living host.

Vitamin D

A chemical (strictly a hormone, despite being called a vitamin) that is crucial to the body's balance of calcium and phosphate in the blood, maintaining strong and healthy bones.

Vitiligo

A disease in which the skin loses its pigmentation in well-demarcated patches. The exact cause is unknown, but it most likely involves an aberrant immune system, which destroys melanocytes (pigment cells) in the skin. Treatment is difficult; some of the many treatment options include camouflage creams, topical steroids, UV light therapy and skin grafting.

Vitreoscillia filiformis

Colourless bacteria (*vitreus* being the Latin for 'transparent') isolated from spa waters. These thin, filamentous bacteria move by gliding over surfaces.

Wolbachia

A group of bacteria that infect insects and parasitic worms. These live within the causative parasitic worms of onchocerciasis (river blindness) and lymphatic filariasis (elephantiasis). Scientists are currently trying to infect mosquitos with this bacterium, as it prevents the dengue virus – responsible for the tropical disease dengue fever – from replicating inside them. The aim is for these mosquitos to mate, spreading the *Wolbachia* bacteria among the population, and eventually rendering wild mosquito populations incapable of spreading dengue fever.

References

Author's Note: Discretion and Definitions

[1] Edelstein, L., 'The Hippocratic Oath: text, translation and interpretation', *Ancient Medicine: Selected Papers of Ludwig Edelstein*, 1943, pp.3–63

1 The Swiss Army Organ

[1] Waring, J. I., 'Early mention of a harlequin fetus in America', *American Journal of Diseases of Children*, 43(2), 1932, p.442

[2] Hovnanian, A., 'Harlequin ichthyosis unmasked: a defect of lipid transport', *The Journal of Clinical Investigation*, 115(7), 2005, pp.1708–10

[3] Rajpopat, S., Moss, C., Mellerio, J., Vahlquist, A., Gånemo, A., Hellstrom-Pigg, M., Ilchyshyn, A., Burrows, N., Lestringant, G., Taylor, A. and Kennedy, C., 'Harlequin ichthyosis: a review of clinical and molecular findings in 45 cases', *Archives of Dermatology*, 147(6), 2011, pp.681–6

[4] Griffiths, C., Barker, J., Bleiker, T., Chalmers, R. and Creamer, D. (eds), *Rook's Textbook of Dermatology*, Vols 1–4, 2016, John Wiley & Sons

[5] Layton, D. W. and Beamer, P. I., 'Migration of contaminated soil and airborne particulates to indoor dust', *Environmental Science & Technology*, 43(21), 2009, pp.8199–205

[6] Weaire, D., 'Kelvin's foam structure: a commentary', *Philosophical Magazine Letters*, 88(2), 2008, pp.91–102

[7] Yokouchi, M., Atsugi, T., Van Logtestijn, M., Tanaka, R. J., Kajimura, M., Suematsu, M., Furuse, M., Amagai, M. and Kubo, A., 'Epidermal cell turnover across tight junctions based on Kelvin's tetrakaidecahedron cell shape', *Elife*, 5, 2016

[8] Hwang, S. and Schwartz, R. A., 'Keratosis pilaris: a common follicular hyperkeratosis', *Cutis*, 82(3), 2008, pp.177–80

[9] Hanifin, J. M., Reed, M. L. and Eczema Prevalance and Impact Working Group, 'A population-based survey of eczema prevalence in the United States', *Dermatitis*, 18(2), 2007, pp.82–91

[10] Maintz, L. and Novak, N., 'Getting more and more complex: the pathophysiology of atopic eczema', *European Journal of Dermatology*, 17(4), 2007, pp.267–83

[11] Palmer, C. N., Irvine, A. D., Terron-Kwiatkowski, A., Zhao, Y., Liao, H., Lee, S. P., Goudie, D. R., Sandilands, A., Campbell, L. E., Smith, F. J. and O'Regan, G. M., 'Common loss-of-function variants of the epidermal barrier protein filaggrin are a major predisposing factor for atopic dermatitis', *Nature Genetics*, 38(4), 2006

[12] Engebretsen, K. A., Kezic, S., Riethmüller, C., Franz, J., Jakasa, I., Hedengran, A., Linneberg, A., Johansen, J. D. and Thyssen, J. P., 'Changes in filaggrin degradation products and corneocyte surface texture by season', *British Journal of Dermatology*, 178(5), 2018, pp.1143–50

[13] Janich, P., Toufighi, K., Solanas, G., Luis, N. M., Minkwitz, S., Serrano, L., Lehner, B. and Benitah, S. A., 'Human epidermal stem cell function is regulated by circadian oscillations', *Cell Stem Cell*, 13(6), 2013, pp.745–53

[14] Wang, H., van Spyk, E., Liu, Q., Geyfman, M., Salmans, M. L., Kumar, V., Ihler, A., Li, N., Takahashi, J. S. and Andersen, B., 'Time-restricted feeding shifts the skin circadian clock and alters UVB-induced DNA damage', *Cell Reports*, 20(5), 2017, pp.1061–72

[15] Hofer, M. K., Collins, H. K., Whillans, A. V. and Chen, F. S., 'Olfactory cues from romantic partners and strangers influence women's responses to stress', *Journal of Personality and Social Psychology*, 114(1), 2018, p.1

[16] Miller, S. L. and Maner, J. K., 'Scent of a woman: Men's testosterone responses to olfactory ovulation cues', *Psychological Science*, 21(2), 2010, pp.276–83

[17] Wedekind, C., Seebeck, T., Bettens, F. and Paepke, A. J., 'MHC-dependent mate preferences in humans', *Proceedings of the Royal Society of London, Series B, Biological Sciences*, 260(1359), 1995, pp.245–9

[18] Kromer, J., Hummel, T., Pietrowski, D., Giani, A. S., Sauter, J., Ehninger, G., Schmidt, A. H. and Croy, I., 'Influence of HLA on human partnership and sexual satisfaction', *Scientific Reports*, 6, 2016, p.32550

[19] Cowburn, A. S., Macias, D., Summers, C., Chilvers, E. R. and Johnson, R. S., 'Cardiovascular adaptation to hypoxia and the role of peripheral resistance', *eLife*, 6, 2017

[20] Carretero, O. A. and Oparil, S., 'Essential hypertension: part I: definition and etiology', *Circulation*, 101(3), 2000, pp.329–35

[21] Langerhans P., 'Über die Nerven der menschlichen Haut', *Archiv für pathologische Anatomie und Physiologie und für klinische Medicin*, 44(2–3), 1868, pp.325–37

[22] Pasparakis, M., Haase, I. and Nestle, F. O., 'Mechanisms regulating skin immunity and inflammation', *Nature Reviews Immunology*, 14(5), 2014, pp.289–301

[23] Mlynek, A., Vieira dos Santos, R., Ardelean, E., Weller, K., Magerl, M., Church, M. K. and Maurer, M., 'A novel, simple, validated and reproducible instrument for assessing provocation threshold levels in patients with symptomatic dermographism', *Clinical and Experimental Dermatology*, 38(4), 2013, pp.60–6

[24] Salimi, M., Barlow, J. L., Saunders, S. P., Xue, L., Gutowska-Owsiak, D., Wang, X., Huang, L. C., Johnson, D., Scanlon, S. T., McKenzie, A. N. and Fallon, P. G., and Ogg, G., 'A role for IL-25 and IL-33–driven type-2 innate lymphoid cells in atopic dermatitis', *Journal of Experimental Medicine*, 210(13), 2013, pp.2939–50

[25] Jabbar-Lopez, Z. K., Yiu, Z. Z., Ward, V., Exton, L. S., Mustapa, M. F. M., Samarasekera, E., Burden, A. D., Murphy, R., Owen, C. M., Parslew, R. and Venning, V., 'Quantitative evaluation of biologic therapy options for psoriasis: a systematic review and network meta-analysis', *Journal of Investigative Dermatology*, 137(8), 2017, pp.1646–54

[26] Warman, P. H. and Ennos, A. R., 'Fingerprints are unlikely to increase the friction of primate fingerpads', *Journal of Experimental Biology*, 212(13), 2009, pp.2016–22

[27] Hirsch, T., Rothoeft, T., Teig, N., Bauer, J. W., Pellegrini, G., De Rosa, L., Scaglione, D., Reichelt, J., Klausegger, A., Kneisz, D. and Romano, O., 'Regeneration of the entire human epidermis using transgenic stem cells', *Nature*, 551(7680), 2017, pp.327–32

2 Skin Safari

[1] Grice, E. A., Kong, H. H., Conlan, S., Deming, C. B., Davis, J., Young, A. C., Bouffard, G. G., Blakesley, R. W., Murray, P. R., Green, E. D. and Turner, M. L., 'Topographical and temporal diversity of the human skin microbiome', *Science*, 324(5931), 2009, pp.1190–92

[2] Human Microbiome Project Consortium, 'Structure, function and diversity of the healthy human microbiome', *Nature*, 486(7402), 2012, pp.207–14

[3] Sender, R., Fuchs, S. and Milo, R., 'Are we really vastly outnumbered? Revisiting the ratio of bacterial to host cells in humans', *Cell*, 164(3), 2016, pp.337–40

[4] Sender, R., Fuchs, S. and Milo, R., 'Revised estimates for the number of human and bacteria cells in the body', *Public Library of Science, Biology*, 14(8), 2016, p.e1002533

[5] Arsenijevic, V. S. A., Milobratovic, D., Barac, A. M., Vekic, B., Marinkovic, J. and Kostic, V. S., 'A laboratory-based study on patients with Parkinson's disease and seborrheic dermatitis: the presence and density of Malassezia yeasts, their different species and enzymes production', *BMC Dermatology*, 14(1), 2014, p.5

[6] Beylot, C., Auffret, N., Poli, F., Claudel, J. P., Leccia, M. T., Del Giudice, P. and Dreno, B., 'Propionibacterium acnes: an update on its role in the pathogenesis of acne', *Journal of the European Academy of Dermatology and Venereology*, 28(3), 2014, pp.271–8

[7] Campisano, A., Ometto, L., Compant, S., Pancher, M., Antonielli, L., Yousaf, S., Varotto, C., Anfora, G., Pertot, I., Sessitsch, A. and Rota-Stabelli, O., 'Interkingdom transfer of the acne-causing agent, Propionibacterium acnes, from human to grapevine', *Molecular Biology and Evolution*, 31(5), 2014, pp.1059–65

[8] Kobayashi, T., Glatz, M., Horiuchi, K., Kawasaki, H., Akiyama, H., Kaplan, D. H., Kong, H. H., Amagai, M. and Nagao, K., 'Dysbiosis and Staphylococcus aureus colonization drives inflammation in atopic dermatitis', *Immunity*, 42(4), 2015, pp.756–66

[9] Surdel, M. C., Horvath, D. J., Lojek, L. J., Fullen, A. R., Simpson, J., Dutter, B. F., Salleng, K. J., Ford, J. B., Jenkins, J. L., Nagarajan, R. and Teixeira, P. L., 'Antibacterial photosensitization through activation of

coproporphyrinogen oxidase', *Proceedings of the National Academy of Sciences of the United States of America*, 114(32), 2017. pp.e6652–59

[10] Nakatsuji, T., Chen, T. H., Butcher, A. M., Trzoss, L. L., Nam, S. J., Shirakawa, K. T., Zhou, W., Oh, J., Otto, M., Fenical, W. and Gallo, R. L., 'A commensal strain of Staphylococcus epidermidis protects against skin neoplasia', *Science Advances*, 4(2), 2018, p.eaao4502

[11] Doroshenko, N., Tseng, B. S., Howlin, R. P., Deacon, J., Wharton, J. A., Thurner, P. J., Gilmore, B. F., Parsek, M. R. and Stoodley, P., 'Extracellular DNA impedes the transport of vancomycin in Staphylococcus epidermidis biofilms preexposed to subinhibitory concentrations of vancomycin', *Antimicrobial Agents and Chemotherapy*, 58(12), 2014, pp.7273–82

[12] Murdoch, D. R., Corey, G. R., Hoen, B., Miró, J. M., Fowler, V. G., Bayer, A. S., Karchmer, A. W., Olaison, L., Pappas, P. A., Moreillon, P. and Chambers, S. T., 'Clinical presentation, etiology, and outcome of infective endocarditis in the 21st century: the International Collaboration on Endocarditis–Prospective Cohort Study', *Archives of internal medicine*, 169(5), 2009, pp.463–73.

[13] Silver, B., Behrouz, R. and Silliman, S., 'Bacterial endocarditis and cerebrovascular disease', *Current Neurology and Neuroscience Reports*, 16(12), 2016, p.104

[14] Blöchl, E., Rachel, R., Burggraf, S., Hafenbradl, D., Jannasch, H. W. and Stetter, K. O., 'Pyrolobus fumarii, gen. and sp. nov., represents a novel group of archaea, extending the upper temperature limit for life to 113 degrees C', *Extremophiles*, 1(1), 1997, pp.14–21

[15] Moissl-Eichinger, C., Probst, A. J., Birarda, G., Auerbach, A., Koskinen, K., Wolf, P. and Holman, H. Y. N., 'Human age and skin physiology shape diversity and abundance of Archaea on skin', *Scientific Reports*, 7(1), 2017, article 4039

[16] Turgut Erdemir, A., Gurel, M. S., Koku Aksu, A. E., Falay, T., Inan Yuksel, E. and Sarikaya, E., 'Demodex mites in acne rosacea: reflectance confocal microscopic study', *Australasian Journal of Dermatology*, 58(2), 2017

[17] Palopoli, M. F., Fergus, D. J., Minot, S., Pei, D.T., Simison, W. B., Fernandez-Silva, I., Thoemmes, M. S., Dunn, R. R. and Trautwein, M., 'Global divergence of the human follicle mite Demodex folliculorum: Persistent associations between host ancestry and mite lineages',

Proceedings of the National Academy of Sciences of the United States of America, 112(52), 2015, pp.15958–63

[18] Roberts, R. J., 'Head lice', *New England Journal of Medicine*, 346(21), 2002, pp.1645–50

[19] Gellatly, K. J., Krim, S., Palenchar, D. J., Shepherd, K., Yoon, K. S., Rhodes, C. J., Lee, S. H. and Marshall Clark, J., 'Expansion of the knockdown resistance frequency map for human head lice (Phthiraptera: Pediculidae) in the United States using quantitative sequencing', *Journal of Medical Entomology*, 53(3), 2016, pp.653–9

[20] Rozsa, L. and Apari, P., 'Why infest the loved ones – inherent human behaviour indicates former mutualism with head lice', *Parasitology*, 139(6), 2012, pp.696–700

[21] Olds, B. P., Coates, B. S., Steele, L. D., Sun, W., Agunbiade, T. A., Yoon, K. S., Strycharz, J. P., Lee, S. H., Paige, K. N., Clark, J. M. and Pittendrigh, B. R., 'Comparison of the transcriptional profiles of head and body lice', *Insect Molecular Biology*, 21(2), 2012, pp.257–68

[22] Welford, M. and Bossak, B., 'Body lice, yersinia pestis orientalis, and black death', *Emerging Infectious Diseases*, 16(10), 2010, p.1649

[23] Armstrong, N. R. and Wilson, J. D., 'Did the "Brazilian" kill the pubic louse?', *Sexually Transmitted Infections*, 82(3), 2006, pp.265–6

[24] Baldo, L., Desjardins, C. A., Russell, J. A., Stahlhut, J. K. and Werren, J. H., 'Accelerated microevolution in an outer membrane protein (OMP) of the intracellular bacteria Wolbachia', *BMC Evolutionary Biology*, 10(1), 2010, p.48

[25] Savioli, L., Daumerie, D. and World Health Organization, 'First WHO report on neglected tropical diseases: working to overcome the global impact of neglected tropical diseases', *Geneva: World Health Organization*, 2010, pp.1–184

[26] Jarrett, R., Salio, M., Lloyd-Lavery, A., Subramaniam, S., Bourgeois, E., Archer, C., Cheung, K. L., Hardman, C., Chandler, D., Salimi, M., Gutowska-Owsiak, D., Bernadino de la Serna, J., Fallon, P. G., Jolin, H., Mckenzie, A,. Dziembowski, A., Podobas, E. I., Bal, W., Johnson, J., Moody, D. B., Cerundolo, V., and Ogg, G., 'Filaggrin inhibits generation of CD1a neolipid antigens by house dust mite-derived phospholipase', *Science Translational Medicine*, 8(325), 2016, p.325ra18

[27] Singh, K., Davies, G., Alenazi, Y., Eaton, J. R., Kawamura, A. and Bhattacharya, S., 'Yeast surface display identifies a family of evasins from ticks with novel polyvalent CC chemokine-binding activities', *Scientific Reports*, 7(1), 2017, article 4267

[28] Szabó, K., Erdei, L., Bolla, B. S., Tax, G., Bíró, T. and Kemény, L., 'Factors shaping the composition of the cutaneous microbiota', *British Journal of Dermatology*, 176(2), 2017, pp.344–51

[29] Haahr, T., Glavind, J., Axelsson, P., Bistrup Fischer, M., Bjurström, J., Andrésdóttir, G., Teilmann-Jørgensen, D., Bonde, U., Olsén Sørensen, N., Møller, M. and Fuglsang, J., 'Vaginal seeding or vaginal microbial transfer from the mother to the caesarean-born neonate: a commentary regarding clinical management', *BJOG: An International Journal of Obstetrics and Gynaecology*, 125(5), 2018, pp.533–6

[30] Cunnington, A. J., Sim, K.., Deierl, A., Kroll, J. S., Brannigan, E. and Darby, J., 'Vaginal seeding of infants born by caesarean section', *British Medical Journal*, 2016, p.i227

[31] Mueller, N. T., Bakacs, E., Combellick, J., Grigoryan, Z. and Dominguez-Bello, M. G., 'The infant microbiome development: mom matters', *Trends in molecular medicine*, 21(2), 2015, pp.109–117

[32] Oh, J., Freeman, A. F., Park, M., Sokolic, R., Candotti, F., Holland, S. M., Segre, J. A., Kong, H. H. and NISC Comparative Sequencing Program, 'The altered landscape of the human skin microbiome in patients with primary immunodeficiencies', *Genome Research*, 23(12), 2013, pp.2103–14

[33] Oh, J., Byrd, A. L., Park, M., Kong, H. H., Segre, J. A. and NISC Comparative Sequencing Program, 'Temporal stability of the human skin microbiome', *Cell*, 165(4), 2016, pp.854–66

[34] Meadow, J. F., Bateman, A. C., Herkert, K. M., O'Connor, T. K. and Green, J. L., 'Significant changes in the skin microbiome mediated by the sport of roller derby', *PeerJ – Life and Environment*, 1, 2013. p.e53

[35] Abeles, S. R., Jones, M. B., Santiago-Rodriguez, T. M., Ly, M., Klitgord, N., Yooseph, S., Nelson, K. E. and Pride, D. T., 'Microbial diversity in individuals and their household contacts following typical antibiotic courses', *Microbiome*, 4(1), 2016, p.39

[36] Ross, A. A., Doxey, A. C. and Neufeld, J. D., 'The skin microbiome of cohabiting couples', *MSystems*, 2(4), 2017, pp.e00043-17

[37] Chase, J., Fouquier, J., Zare, M., Sonderegger, D. L., Knight, R., Kelley, S. T., Siegel, J. and Caporaso, J. G., 'Geography and location are the primary drivers of office microbiome composition', *MSystems*, 1(2), 2016, pp.e00022-16

[38] Gimblet, C., Meisel, J. S., Loesche, M. A., Cole, S. D., Horwinski, J., Novais, F. O., Misic, A. M., Bradley, C. W., Beiting, D. P., Rankin, S. C. and Carvalho, L. P., 'Cutaneous Leishmaniasis induces a transmissible dysbiotic skin microbiota that promotes skin inflammation', *Cell Host & Microbe*, 22(1), 2017, pp.13–24

[39] Scharschmidt, T. C., Vasquez, K. S., Truong, H. A., Gearty, S. V., Pauli, M. L., Nosbaum, A., Gratz, I. K., Otto, M., Moon, J. J., Liese, J. and Abbas, A. K., 'A wave of regulatory T cells into neonatal skin mediates tolerance to commensal microbes', *Immunity*, 43(5), 2015, pp.1011–21

[40] Lambrecht, B. N. and Hammad, H., 'The immunology of the allergy epidemic and the hygiene hypothesis', *Nature Immunology*, 18(10), 2017, pp.1076–83

[41] Volz, T., Skabytska, Y., Guenova, E., Chen, K. M., Frick, J. S., Kirschning, C. J., Kaesler, S., Röcken, M. and Biedermann, T., 'Nonpathogenic bacteria alleviating atopic dermatitis inflammation induce IL-10-producing dendritic cells and regulatory Tr1 cells', *Journal of Investigative Dermatology*, 134(1), 2014, pp.96–104

[42] Kassam, Z., Lee, C. H., Yuan, Y. and Hunt, R. H., 'Fecal microbiota transplantation for Clostridium difficile infection: systematic review and meta-analysis', *The American Journal of Gastroenterology*, 108(4), 2013, p.500

[43] Jeong, J. H., Lee, C. Y. and Chung, D. K., 2016. 'Probiotic lactic acid bacteria and skin health', *Critical Reviews in Food Science and Nutrition*, 56(14), pp.2331–7

[44] Holz, C., Benning, J., Schaudt, M., Heilmann, A., Schultchen, J., Goelling, D. and Lang, C., 'Novel bioactive from Lactobacillus brevis DSM17250 to stimulate the growth of Staphylococcus epidermidis: a pilot study', *Beneficial Microbes*, 8(1), 2017, pp.121–31

[45] Coughlin, C. C., Swink, S. M., Horwinski, J., Sfyroera, G., Bugayev, J., Grice, E. A. and Yan, A. C., 'The preadolescent acne microbiome: A prospective, randomized, pilot study investigating characterization and effects of acne therapy', *Pediatric Dermatology*, 34(6), 2017, pp.661–4

[46] Callewaert, C., Kerckhof, F. M., Granitsiotis, M. S., Van Gele, M., Van de Wiele, T. and Boon, N., 'Characterization of Staphylococcus and Corynebacterium clusters in the human axillary region', *PLOS ONE*, 8(8), 2013, p.e70538

[47] Callewaert, C., Lambert, J. and Van de Wiele, T., 'Towards a bacterial treatment for armpit malodour', *Experimental Dermatology*, 26(5), 2017, pp.388–91

3 Gut Feeling

[1] Çerman, A. A., Aktaş, E., Altunay, İ. K., Arıcı, J. E., Tulunay, A. and Ozturk, F. Y., 'Dietary glycemic factors, insulin resistance, and adiponectin levels in acne vulgaris', *Journal of the American Academy of Dermatology*, 75(1), 2016, pp.155–62

[2] Smith, R. N., Mann, N. J., Braue, A., Mäkeläinen, H. and Varigos, G. A., 'A low-glycemic-load diet improves symptoms in acne vulgaris patients: a randomized controlled trial', *The American Journal of Clinical Nutrition*, 86(1), 2007, p.107–15

[3] Williams, S. in 'How the derms do it: 4 expert dermatologists on their daily skincare routines', *Get the Gloss*, 10 November 2017

[4] Fulton, J. E., Plewig, G., Kligman, A. M., 'Effect of Chocolate on Acne Vulgaris', *JAMA Network*, 210(11), 1969, pp.2071–4

[5] Davidovici, B. B. and Wolf, R., 'The role of diet in acne: facts and controversies', *Clinics in Dermatology*, 28(1), 2010, pp.12–16

[6] Caperton, C., Block, S., Viera, M., Keri, J. and Berman, B., 'Double-blind, placebo-controlled study assessing the effect of chocolate consumption in subjects with a history of acne vulgaris', *The Journal of Clinical and Aesthetic Dermatology*, 7(5), 2014, p.19

[7] Fialová, J., Roberts, S. C. and Havlíček, J., 'Consumption of garlic positively affects hedonic perception of axillary body odour', *Appetite*, 97, 2016, pp.8–15

[8] Havlicek, J. and Lenochova, P., 'The effect of meat consumption on body odor attractiveness', *Chemical senses*, 31(8), 2006, pp.747–52

[9] Bronsnick, T., Murzaku, E. C. and Rao, B. K., 'Diet in dermatology: Part I. Atopic dermatitis, acne, and nonmelanoma skin cancer', *Journal of the American Academy of Dermatology*, 71(6), 2014, p.1039

[10] Clarke, K. A., Dew, T. P., Watson, R. E., Farrar, M. D., Osman, J. E., Nicolaou, A., Rhodes, L. E. and Williamson, G., 'Green tea catechins and their metabolites in human skin before and after exposure to ultraviolet radiation', *The Journal of Nutritional Biochemistry*, 27, 2016, pp.203–10

[11] Moon, T. E., Levine, N., Cartmel, B., Bangert, J. L., Rodney, S., Dong, Q., Peng, Y. M. and Alberts, D. S., 'Effect of retinol in preventing squamous cell skin cancer in moderate-risk subjects: a randomized, double-blind, controlled trial. Southwest Skin Cancer Prevention Study Group', *Cancer Epidemiology and Prevention Biomarkers*, 6(11), 1997, pp.949–56

[12] Cooperstone, J. L., Tober, K. L., Riedl, K. M., Teegarden, M. D., Cichon, M. J., Francis, D. M., Schwartz, S. J. and Oberyszyn, T. M., 'Tomatoes protect against development of UV-induced keratinocyte carcinoma via metabolomic alterations', *Scientific Reports*, 7(1), 2017, article 5106

[13] Foo, Y. Z., Rhodes, G. and Simmons, L. W., 'The carotenoid beta-carotene enhances facial color, attractiveness and perceived health, but not actual health, in humans', *Behavioral Ecology*, 28(2), 2017, pp.570–78

[14] Lefevre, C. E. and Perrett, D. I., 'Fruit over sunbed: carotenoid skin colouration is found more attractive than melanin colouration', *The Quarterly Journal of Experimental Psychology*, 68(2), 2015, pp.284–93

[15] Stephen, I. D., Coetzee, V. and Perrett, D. I., 'Carotenoid and melanin pigment coloration affect perceived human health', *Evolution and Human Behavior*, 32(3), 2011, pp.216–27

[16] Watson, J., 2013. 'Oxidants, antioxidants and the current incurability of metastatic cancers', *Open Biology*, 3(1), p.120144

[17] Sidbury, R., Tom, W. L., Bergman, J. N., Cooper, K. D., Silverman, R. A., Berger, T. G., Chamlin, S. L., Cohen, D. E., Cordoro, K. M., Davis, D. M. and Feldman, S. R., 'Guidelines of care for the management of atopic dermatitis: Section 4. Prevention of disease flares and use of adjunctive therapies and approaches', *Journal of the American Academy of Dermatology*, 71(6), 2014, pp.1218–33

[18] Hata, T. R., Audish, D., Kotol, P., Coda, A., Kabigting, F., Miller, J., Alexandrescu, D., Boguniewicz, M., Taylor, P., Aertker, L. and Kesler, K., 'A randomized controlled double-blind investigation of the effects of vitamin D dietary supplementation in subjects with atopic dermatitis', *Journal of The European Academy of Dermatology and Venereology*, 28(6), 2014, pp.781–9

[19] Amestejani, M., Salehi, B. S., Vasigh, M., Sobhkhiz, A., Karami, M., Alinia, H., Kamrava, S. K., Shamspour, N., Ghalehbaghi, B. and Behzadi, A. H., 'Vitamin D supplementation in the treatment of atopic dermatitis: a clinical trial study', *Journal of Drugs in Dermatology*, 11(3), 2012, pp.327–30

[20] Ma, C. A., Stinson, J. R., Zhang, Y., Abbott, J. K., Weinreich, M. A., Hauk, P. J., Reynolds, P. R., Lyons, J. J., Nelson, C. G., Ruffo, E. and Dorjbal, B., 'Germline hypomorphic CARD11 mutations in severe atopic disease', *Nature Genetics*, 49(8), 2017, p.1192

[21] Jensen, P., Zachariae, C., Christensen, R., Geiker, N. R., Schaadt, B. K., Stender, S., Hansen, P. R., Astrup, A. and Skov, L., 'Effect of weight loss on the severity of psoriasis: a randomized clinical study', *JAMA Dermatology*, 149(7), 2013, pp.795–801

[22] Singh, S., Sonkar, G. K. and Singh, S., 'Celiac disease-associated antibodies in patients with psoriasis and correlation with HLA Cw6', *Journal of Clinical Laboratory Analysis*, 24(4), 2010, pp.269–72

[23] Wolf, R., Wolf, D., Rudikoff, D. and Parish, L. C., 'Nutrition and water: drinking eight glasses of water a day ensures proper skin hydration – myth or reality?', *Clinics in Dermatology*, 28(4), 2010, pp.380–83

[24] Negoianu, D. and Goldfarb, S., 'Just add water', *Journal of the American Society of Nephrology*, 19(6), 2008, pp.1041–3

[25] Rota, M., Pasquali, E., Bellocco, R., Bagnardi, V., Scotti, L., Islami, F., Negri, E., Boffetta, P., Pelucchi, C., Corrao, G. and La Vecchia, C., 'Alcohol drinking and cutaneous melanoma risk: a systematic review and dose–risk meta-analysis', *British Journal of Dermatology*, 170(5), 2014, pp.1021–28

[26] Transparency Market Research, 'Nutricosmetics Market – Global Industry Analysis, Size, Share, Growth, Trends and Forecast 2014–2020', 2015

[27] Borumand, M. and Sibilla, S., 'Effects of a nutritional supplement containing collagen peptides on skin elasticity, hydration and wrinkles', *Journal of Medical Nutrition and Nutraceuticals*, 4(1), 2015, pp.47–53

[28] Borumand, M. and Sibilla, S., 'Daily consumption of the collagen supplement Pure Gold Collagen® reduces visible signs of aging', *Clinical Interventions in Aging*, 9, 2014, p.1747

[29] Etheridge, E. W., *The Butterfly Caste: A Social History of Pellagra in the South*, Greenwood, 1972

[30] Clay, K., Schmick, E. and Troesken, W., 'The Rise and Fall of Pellagra in the American South', *National Bureau of Economic Research*, 2017, p.w23730

[31] Werfel, T., Heratizadeh, A., Aberer, W., Ahrens, F., Augustin, M., Biedermann, T., Diepgen, T., Fölster-Holst, R., Gieler, U., Kahle, J. and Kapp, A., 'S2k guideline on diagnosis and treatment of atopic dermatitis – short version', *Allergo Journal International*, 25(3), 2016, pp.82–95

[32] Zuberbier, T., Aberer, W., Asero, R., Bindslev-Jensen, C., Brzoza, Z., Canonica, G. W., Church, M. K., Ensina, L. F., Giménez-Arnau, A., Godse, K. and Gonçalo, M., 'The EAACI/GA(2) LEN/EDF/WAO Guideline for the definition, classification, diagnosis, and management of urticaria: the 2013 revision and update', *Allergy*, 69(7), 2014, pp.868–87

[33] Zuberbier, T., Chantraine-Hess, S., Hartmann, K. and Czarnetzki, B. M., 'Pseudoallergen-free diet in the treatment of chronic urticaria. A prospective study', *Acta Dermato-venereologica*, 75(6), 1995, pp.484–7

[34] Parodi, A., Paolino, S., Greco, A., Drago, F., Mansi, C., Rebora, A., Parodi, A. and Savarino, V., 'Small intestinal bacterial overgrowth in rosacea: clinical effectiveness of its eradication', *Clinical Gastroenterology and Hepatology*, 6(7), 2008, pp.759–64

[35] Jeong, J. H., Lee, C. Y. and Chung, D. K., 'Probiotic lactic acid bacteria and skin health', *Critical Reviews in Food Science and Nutrition*, 56(14), 2016, pp.2331–7

[36] Meneghin, F., Fabiano, V., Mameli, C. and Zuccotti, G. V., 'Probiotics and atopic dermatitis in children', *Pharmaceuticals*, 5(7), 2012, pp.727–44

[37] Chang, Y. S., Trivedi, M. K., Jha, A., Lin, Y. F., Dimaano, L. and García-Romero, M. T., 'Synbiotics for prevention and treatment of atopic dermatitis: a meta-analysis of randomized clinical trials', *JAMA Pediatrics*, 170(3), 2016, pp.236–42

[38] Smits, H. H., Engering, A., van der Kleij, D., de Jong, E. C., Schipper, K., van Capel, T. M., Zaat, B. A., Yazdanbakhsh, M., Wierenga, E. A., van Kooyk, Y. and Kapsenberg, M. L., 'Selective probiotic bacteria induce IL-10-producing regulatory T cells in vitro by modulating dendritic cell function through dendritic cell-specific intercellular adhesion molecule 3-grabbing nonintegrin', *Journal of Allergy and Clinical Immunology*, 115(6), 2005, pp.1260–7

[39] O'Neill, C.A., Monteleone, G., McLaughlin, J. T. and Paus, R., 'The gut–skin axis in health and disease: A paradigm with therapeutic implications', *BioEssays*, 38(11), 2016, pp.1167–76

[40] Zákostelská, Z., Málková, J., Klimešová, K., Rossmann, P., Hornová, M., Novosádová, I., Stehlíková, Z., Kostovčík, M., Hudcovic, T., Štepánková, R. and Jůzlová, K., 'Intestinal microbiota promotes psoriasis-like skin inflammation by enhancing Th17 response', *PLOS ONE*, 11(7), 2016, p.e0159539

[41] Zanvit, P., Konkel, J. E., Jiao, X., Kasagi, S., Zhang, D., Wu, R., Chia, C., Ajami, N. J., Smith, D. P., Petrosino, J. F. and Abbatiello, B., 'Antibiotics in neonatal life increase murine susceptibility to experimental psoriasis', *Nature Communications*, 6, 2015

[42] Plantamura, E., Dzutsev, A., Chamaillard, M., Djebali, S., Moudombi, L., Boucinha, L., Grau, M., Macari, C., Bauché, D., Dumitrescu, O. and Rasigade, J. P., 'MAVS deficiency induces gut dysbiotic microbiota conferring a proallergic phenotype', *Proceedings of the National Academy of Sciences of the United States of America*, 115(41), 2018, pp.10404–9

[43] Stokes, J. H. and Pillsbury, D. M., 'The effect on the skin of emotional and nervous states. III: Theoretical and practical consideration of a gastro-intestinal mechanism', *Archives of Dermatology and Syphilology*, 22(6), 1930, pp.962–93

[44] Kelly, J. R., Kennedy, P. J., Cryan, J. F., Dinan, T. G., Clarke, G. and Hyland, N. P., 'Breaking down the barriers: the gut microbiome, intestinal permeability and stress-related psychiatric disorders', *Frontiers in Cellular Neuroscience*, 9, 2015, p.392

[45] Bailey, M. T., Dowd, S. E., Galley, J. D., Hufnagle, A. R., Allen, R. G. and Lyte, M., 'Exposure to a social stressor alters the structure of the intestinal microbiota: implications for stressor-induced immunomodulation', *Brain, Behavior, and Immunity*, 25(3), 2011, pp.397–407

[46] Savignac, H. M., Kiely, B., Dinan, T. G. and Cryan, J. F., 'Bifidobacteria exert strain-specific effects on stress-related behavior and physiology in BALB/c mice', *Neurogastroenterology & Motility*, 26(11), 2014, pp.1615–27

[47] Kelly, J. R., Kennedy, P. J., Cryan, J. F., Dinan, T. G., Clarke, G. and Hyland, N. P., 'Breaking down the barriers: the gut microbiome, intestinal

permeability and stress-related psychiatric disorders', *Frontiers in Cellular Neuroscience*, 9, 2015

[48] Du Toit, G., Roberts, G., Sayre, P. H., Plaut, M., Bahnson, H. T., Mitchell, H., Radulovic, S., Chan, S., Fox, A., Turcanu, V. and Lack, G., 'Identifying infants at high risk of peanut allergy: the Learning Early About Peanut Allergy (LEAP) screening study,' *The Journal of Allergy and Clinical Immunology*, 131(1), 2013, pp.135–43

[49] Kelleher, M. M., Dunn-Galvin, A., Gray, C., Murray, D. M., Kiely, M., Kenny, L., McLean, W. I., Irvine, A. D. and Hourihane, J. O. B., 'Skin barrier impairment at birth predicts food allergy at 2 years of age', *The Journal of Allergy and Clinical Immunology*, 137(4), 2016, pp.1111–6

[50] Flohr, C., Perkin, M., Logan, K., Marrs, T., Radulovic, S., Campbell, L. E., MacCallum, S. F., McLean, W. I. and Lack, G., 'Atopic dermatitis and disease severity are the main risk factors for food sensitization in exclusively breastfed infants', *Journal of Investigative Dermatology*, 134(2), 2014, pp.345–50

[51] Walker, M. T., Green, J. E., Ferrie, R. P., Queener, A. M., Kaplan, M. H. and Cook-Mills, J. M., 'Mechanism for initiation of food allergy: dependence on skin barrier mutations and environmental allergen costimulation', *Journal of Allergy and Clinical Immunology*, 141(5), 2018, pp.1711–25

4 Towards the Light

[1] Driver, S. P., Andrews, S. K., Davies, L. J., Robotham, A. S., Wright, A. H., Windhorst, R. A., Cohen, S., Emig, K., Jansen, R. A. and Dunne, L., 'Measurements of extragalactic background light from the far UV to the Far IR from deep ground- and space-based galaxy counts', *The Astrophysical Journal*, 827(2), 2016, p.108

[2] Corani, A., Huijser, A., Gustavsson, T., Markovitsi, D., Malmqvist, P. Å., Pezzella, A., d'Ischia, M. and Sundström, V., 'Superior photoprotective motifs and mechanisms in eumelanins uncovered', *Journal of the American Chemical Society*, 136(33), 2014, pp.11626–35

[3] Dennis, L. K., Vanbeek, M. J., Freeman. L. E. B., Smith, B. J., Dawson, D. V. and Coughlin, J. A., 'Sunburns and risk of cutaneous melanoma: does age matter? A comprehensive meta-analysis', *Annals of Epidemiology*, 18(8), 2008, pp.614–27.

[4] Wu, S., Han, J., Laden, F. and Qureshi, A. A., 'Long-term ultraviolet flux, other potential risk factors, and skin cancer risk: a cohort study', *Cancer Epidemiology and Prevention Biomarkers*, 23(6), 2014, pp.1080–9

[5] Guy, G. P. Jnr, Machlin, S. R., Ekwueme, D. U. and Yabroff, K. R., 'Prevalence and costs of skin cancer treatment in the US, 2002–2006 and 2007–2011', *American Journal of Preventive Medicine*, 48(2), 2015, pp.183–7

[6] Australian Institute of Health and Welfare & Australasian Association of Cancer, 'Cancer in Australia: in brief 2017', Cancer series no. 102. Cat. no. CAN 101.

[7] Muzic, J. G., Schmitt, A. R., Wright, A. C., Alniemi, D. T., Zubair, A. S., Lourido, J. M. O., Seda, I. M. S., Weaver, A. L. and Baum, C. L., 'Incidence and trends of basal cell carcinoma and cutaneous squamous cell carcinoma: a population-based study in Olmsted County, Minnesota, 2000 to 2010', *Mayo Clinic Proceedings*, 92(6), 2017, pp.890–8

[8] Karimkhani, C., Green, A. C., Nijsten, T., Weinstock, M. A., Dellavalle, R. P., Naghavi, M. and Fitzmaurice, C., 'The global burden of melanoma: results from the Global Burden of Disease Study 2015', *British Journal of Dermatology*, 177(1), 2017, pp.134–40

[9] Smittenaar, C. R., Petersen, K. A., Stewart, K., Moitt, N., 'Cancer incidence and mortality projections in the UK until 2035', *British Journal of Cancer*, 115, 2016, pp.1147–55

[10] Conic, R. Z., Cabrera, C. I., Khorana, A. A. and Gastman, B. R., 'Determination of the impact of melanoma surgical timing on survival using the National Cancer Database', *Journal of the American Academy of Dermatology*, 78(1), 2018, pp.40–46

[11] Cymerman, R. M., Wang, K., Murzaku, E. C., Penn, L. A., Osman, I., Shao, Y. and Polsky, D., 'De novo versus nevus-associated melanomas: differences in associations with prognostic indicators and survival', *American Society of Clinical Oncology*, 2015

[12] Dinnes, J., Deeks, J. J., Grainge, M. J., Chuchu, N., di Ruffano, L. F., Matin, R. N., Thomson, D. R., Wong, K. Y., Aldridge, R. B., Abbott, R. and Fawzy, M., 'Visual inspection for diagnosing cutaneous melanoma in adults', *Cochrane Database of Systematic Reviews*, 12, 2018

[13] Pathak, M. A., Jimbow, K., Szabo, G. and Fitzpatrick, T. B., 'Sunlight and melanin pigmentation', *Photochemical and Photobiological Reviews*, 1, 1976, pp.211–39

[14] Ljubešic, N. and Fišer, D., 'A global analysis of emoji usage', *Proceedings of the 10th Web As Corpus Workshop, Association for Computational Linguistics*, 2016, p.82

[15] Lyman, M., Mills, J. O. and Shipman, A. R., 'A dermatological questionnaire for general practitioners in England with a focus on melanoma; misdiagnosis in black patients compared to white patients', *Journal of The European Academy of Dermatology and Venereology*, 31(4), 2017, pp.625–8

[16] Royal Pharmaceutical Society press release, 'RPS calls for clearer labelling on sunscreens after survey reveals confusion', 2015

[17] Corbyn, Z., 'Prevention: lessons from a sunburnt country', *Nature*, 515, 2014, pp.S114–6

[18] British Association of Dermatologists, 'Brits burying their heads in the sand over UK's most common cancer, survey finds', *BAD Press Releases*, 4/5/15.

[19] Seité, S., Del Marmol, V., Moyal, D. and Friedman, A. J., 'Public primary and secondary skin cancer prevention, perceptions and knowledge: an international cross-sectional survey', *Journal of the European Academy of Dermatology and Venereology*, 31(5), 2017, pp.815–20.

[20] Fell, G. L., Robinson, K. C., Mao, J., Woolf, C. J. and Fisher, D. E., 'Skin β-endorphin mediates addiction to UV light', *Cell*, 157(7), 2014, pp.1527–34

[21] Pezdirc, K., Hutchesson, M. J., Whitehead, R., Ozakinci, G., Perrett, D. and Collins, C. E., 'Fruit, vegetable and dietary carotenoid intakes explain variation in skin-color in young Caucasian women: a cross-sectional study', *Nutrients*, 7(7), 2015, pp.5800–15

[22] Mujahid, N., Liang, Y., Murakami, R., Choi, H. G., Dobry, A. S., Wang, J., Suita, Y., Weng, Q. Y., Allouche, J., Kemeny, L. V. and Hermann, A. L., 'A UV-independent topical small-molecule approach for melanin production in human skin', *Cell Reports*, 19(11), 2017, pp.2177–84

[23] Cleaver, J. E., 'Common pathways for ultraviolet skin carcinogenesis in the repair and replication defective groups of xeroderma pigmentosum', *Journal of Dermatological Science*, 23(1), 2000, pp.1–11

[24] Cleaver, J. E., 'Defective repair replication of DNA in xeroderma pigmentosum', *Nature*, 218, 1968, pp.652–6

[25] Bailey, L. R., *The Long Walk: A History of the Navajo Wars, 1846–68*, Westernlore Press, 1964

[26] Rashighi, M. and Harris, J. E., 'Vitiligo pathogenesis and emerging treatments', *Dermatologic Clinics*, 35(2), 2017, pp.257–65

[27] Grzybowski, A. and Pietrzak, K., 'From patient to discoverer – Niels Ryberg Finsen (1860–1904) – the founder of phototherapy in dermatology', *Clinics in Dermatology*, 30(4), 2012, pp.451–5

[28] Watts, G., 'Richard John Cremer', *The Lancet*, 383(9931), 2014, p.1800

[29] Lucey, J. F., 'Neonatal jaundice and phototherapy', *Pediatric Clinics of North America*, 19(4), 1972, pp.827–39

[30] Quandt, B. M., Pfister, M. S., Lübben, J. F., Spano, F., Rossi, R. M., Bona, G. L. and Boesel, L. F., 'POF-yarn weaves: controlling the light out-coupling of wearable phototherapy devices', *Biomedical Optics Express*, 8(10), 2017, pp.4316–30

[31] Car, J., Car, M., Hamilton, F., Layton, A., Lyons, C. and Majeed, A., 'Light therapies for acne', *Cochrane Library*, 2009

[32] Ondrusova, K., Fatehi, M., Barr, A., Czarnecka, Z., Long, W., Suzuki, K., Campbell, S., Philippaert, K., Hubert, M., Tredget, E. and Kwan, P., 'Subcutaneous white adipocytes express a light sensitive signaling pathway mediated via a melanopsin/TRPC channel axis', *Scientific Reports*, 7, 2017, article 16332

[33] Mohammad, K. I., Kassab, M., Shaban, I., Creedy, D. K. and Gamble, J., 'Postpartum evaluation of vitamin D among a sample of Jordanian women', *Journal of Obstetrics and Gynaecology*, 37(2), 2017, pp.200–4

[34] Wolpowitz, D. and Gilchrest, B. A., 'The vitamin D questions: how much do you need and how should you get it?', *Journal of the American Academy of Dermatology*, 54(2), 2006, pp.301–17

[35] Petersen, B., Wulf, H. C., Triguero-Mas, M., Philipsen, P. A., Thieden, E., Olsen, P., Heydenreich, J., Dadvand, P., Basagana, X., Liljendahl, T. S. and Harrison, G. I., 'Sun and ski holidays improve vitamin D status, but are associated with high levels of DNA damage', *Journal of Investigative Dermatology*, 134(11), 2014, pp.2806–13

[36] American Academy of Dermatology 2010 Position Statement: https://www.aad.org/Forms/Policies/Uploads/PS/PS-Vitamin%20D%20Position%20Statement.pdf

5 Ageing Skin

[1] Dealey, C., Posnett, J. and Walker, A., 'The cost of pressure ulcers in the United Kingdom', *Journal of Wound Care*, 21(6), 2012

[2] Huxley, A., *Brave New World*, Vintage Classics, 2007

[3] Kaidbey, K. H., Agin, P. P., Sayre, R. M. and Kligman, A. M., 'Photoprotection by melanin – a comparison of black and Caucasian skin', *Journal of the American Academy of Dermatology*, 1(3), 1979, pp.249–60

[4] Zhang, L., Xiang Chen, S., Guerrero-Juarez, G. F., Li, F., Tong, Y., Liang, Y., Liggins, M., Chen, X., Chen, H., Li, M., Hata, T., Zheng, Y., Plikus, M. V., Gallo, R. L., 'Age-related loss of innate immune antimicrobial function of dermal fat is mediated by transforming growth factor beta', *Immunity*, 2018; DOI: 10.1016/j.immuni.2018.11.003

[5] Brennan, M., Bhatti, H., Nerusu, K. C., Bhagavathula, N., Kang, S., Fisher, G. J., Varani, J. and Voorhees, J. J., 'Matrix metalloproteinase-1 is the major collagenolytic enzyme responsible for collagen damage in UV-irradiated human skin', *Photochemistry and Photobiology*, 78(1), 2003, pp.43–8

[6] Liebel, F., Kaur, S., Ruvolo, E., Kollias, N. and Southall, M. D., 'Irradiation of skin with visible light induces reactive oxygen species and matrix-degrading enzymes', *Journal of Investigative Dermatology*, 132(7), 2012, pp.1901–7

[7] Lee, E. J., Kim, J. Y. and Oh, S. H., 'Advanced glycation end products (AGEs) promote melanogenesis through receptor for AGEs', *Scientific Reports*, 6, 2016, article 27848

[8] Morita, A., 'Tobacco smoke causes premature skin aging', *Journal of Dermatological Science*, 48(3), 2007, pp.169–5

[9] Buffet. J., 'Barefoot Children', *Barometer Soup*, Universal Music Catalogue, 2000

[10] Vierkötter, A., Schikowski, T., Ranft, U., Sugiri, D., Matsui, M., Krämer, U. and Krutmann, J., 'Airborne particle exposure and extrinsic skin aging', *Journal of Investigative Dermatology*, 130(12), 2010, pp.2719–26

[11] London Air Quality Network (LAQN), 'London air data from the first week of 2017', King's College London Environmental Research Group, 2017

[12] Jaliman, D., *Skin Rules*, St Martin's Press, 2013

[13] Axelsson, J., Sundelin, T., Ingre, M., Van Someren, E. J., Olsson, A. and Lekander, M., 'Beauty sleep: experimental study on the perceived health and attractiveness of sleep deprived people', *BMJ*, 341, 2010, p.c6614

[14] Sundelin, T., Lekander, M., Kecklund, G., Van Someren, E. J., Olsson, A. and Axelsson, J., 'Cues of fatigue: effects of sleep deprivation on facial appearance', *Sleep*, 36(9), 2013, pp.1355–60

[15] Oyetakin-White, P., Suggs, A., Koo, B., Matsui, M. S., Yarosh, D., Cooper, K. D. and Baron, E. D., 'Does poor sleep quality affect skin ageing?', *Clinical and Experimental Dermatology*, 2015, 40(1), pp.17–22

[16] Danby, S., Study at the University of Sheffield on BBC's *The Truth About . . . Looking Good*, 2018

[17] Kligman, A. M., Mills, O. H., Leyden, J. J., Gross, P. R., Allen, H. B. and Rudolph, R. I., 'Oral vitamin A in acne vulgaris Preliminary report', *International Journal of Dermatology*, 20(4), 1981, pp.278–85

[18] Hornblum, A. M., *Acres of skin: Human Experiments at Holmesburg Prison*, Routledge, 2013

[19] Boudreau, M. D., Beland, F. A., Felton, R. P., Fu, P. P., Howard, P. C., Mellick, P. W., Thorn, B. T. and Olson, G. R., 'Photo-co-carcinogenesis of Topically Applied Retinyl Palmitate in SKH-1 Hairless Mice', *Photochemistry and Photobiology*, 94(4), 2017, pp.1096–114

[20] Wang, S. Q., Dusza, S. W. and Lim, H. W., 'Safety of retinyl palmitate in sunscreens: a critical analysis', *Journal of the American Academy of Dermatology*, 63(5), 2010, pp.903–90

[21] Leslie Baumann in 'Skincare: The Vitamin A Controversy', *youbeauty*, 2011

[22] Jones, R. R., Castelletto, V., Connon, C. J. and Hamley, I. W., 'Collagen stimulating effect of peptide amphiphile C16–KTTKS on human fibroblasts', *Molecular Pharmaceutics*, 10(3), 2013, pp.1063–69

[23] Watson, R. E. B., Ogden, S., Cotterell, L. F., Bowden, J. J., Bastrilles, J. Y., Long, S. P. and Griffiths, C. E. M., 'A cosmetic "anti-ageing" product

improves photoaged skin: a double-blind, randomized controlled trial', *British Journal of Dermatology*, 161(2), 2009, pp.419–26

[24] Van Ermengem, É., 'A new anaerobic bacillus and its relation to botulism', *Reviews of Infectious Diseases*, 1(4), 1979, pp.701–19

[25] Carruthers, J. D. and Carruthers, J. A., 'Treatment of glabellar frown lines with C. botulinum-A exotoxin', *Journal of Dermatologic Surgery and Oncology*, 18(1), 1992, pp.17–21

[26] Yu, B., Kang, S. Y., Akthakul, A., Ramadurai, N., Pilkenton, M., Patel, A., Nashat, A., Anderson, D. G., Sakamoto, F. H., Gilchrest, B. A. and Anderson, R. R., 'An elastic second skin', *Nature Materials*, 15(8), 2016. pp.911–18

6 The First Sense

[1] Abraira, V. E. and Ginty, D. D., 'The sensory neurons of touch', *Neuron*, 79(4), 2013. pp.618–39

[2] Woo, S. H., Ranade, S., Weyer, A. D., Dubin, A. E., Baba, Y., Qiu, Z., Petrus, M., Miyamoto, T., Reddy, K., Lumpkin, E. A. and Stucky, C. L., 'Piezo2 is required for Merkel-cell mechanotransduction', *Nature*, 509, 2014, pp.622–6

[3] Thought experiment inspired by Linden, D. J., *Touch: The Science of Hand, Heart and Mind*, Penguin, 2016

[4] Penfield, W., and Jasper, H., *Epilepsy and the Functional Anatomy of the Human Brain*, Little, Brown, 1954

[5] Cohen, L. G., Celnik, P., Pascual-Leone, A., Corwell, B., Faiz, L., Dambrosia, J., Honda, M., Sadato, N., Gerloff, C., Catalá, M. D. and Hallett, M., 'Functional relevance of cross-modal plasticity in blind humans', *Nature*, 389, 1997, pp.180–83

[6] Ro, T., Farnè, A., Johnson, R. M., Wedeen, V., Chu, Z., Wang, Z. J., Hunter, J. V. and Beauchamp, M. S., 'Feeling sounds after a thalamic lesion', *Annals of Neurology*, 62(5), 2007, pp.433–41

[7] Changizi, M., Weber, R., Kotecha, R. and Palazzo, J., 'Are wet-induced wrinkled fingers primate rain treads?' *Brain, Behavior and Evolution*, 77(4), 2011, pp.286–90

[8] Kareklas, K., Nettle, D. and Smulders, T. V., 'Water-induced finger wrinkles improve handling of wet objects', *Biology Letters*, 9(2), 2013, p.20120999

[9] Haseleu, J., Omerbašić, D., Frenzel, H., Gross, M. and Lewin, G. R., 'Water-induced finger wrinkles do not affect touch acuity or dexterity in handling wet objects', *PLOS ONE*, 9(1), 2014, p.e84949

[10] Hertenstein, M. J., Holmes, R., McCullough, M. and Keltner, D., 'The communication of emotion via touch', *Emotion*, 9(4), 2009, p.566

[11] Liljencrantz, J. and Olausson, H., 'Tactile C fibers and their contributions to pleasant sensations and to tactile allodynia', *Frontiers in Behavioral Neuroscience*, 8, 2014

[12] Brauer, J., Xiao, Y., Poulain, T., Friederici, A. D. and Schirmer, A., 'Frequency of maternal touch predicts resting activity and connectivity of the developing social brain', *Cerebral Cortex*, 26(8), 2016, pp.3544–52

[13] Walker, S. C., Trotter, P. D., Woods, A. and McGlone, F., 'Vicarious ratings of social touch reflect the anatomical distribution & velocity tuning of C-tactile afferents: a hedonic homunculus?', *Behavioural Brain Research*, 320, 2017, pp.91–6

[14] Suvilehto, J. T., Glerean, E., Dunbar, R. I., Hari, R. and Nummenmaa, L., 'Topography of social touching depends on emotional bonds between humans', *Proceedings of the National Academy of Sciences*, 112(45), 2015, pp.13811–6

[15] van Stralen, H. E., van Zandvoort, M. J., Hoppenbrouwers, S. S., Vissers, L. M., Kappelle, L. J. and Dijkerman, H. C., 'Affective touch modulates the rubber hand illusion', *Cognition*, 131(1), 2014, pp.147–58

[16] Blakemore, S. J., Wolpert, D. M. and Frith, C. D., 'Central cancellation of self-produced tickle sensation', *Nature Neuroscience*, 1(7), 1998, pp.635–40

[17] Linden, D. J., *Touch: The Science of Hand, Heart and Mind*, Penguin, 2016

[18] Cox, J. J., Reimann, F., Nicholas, A. K., Thornton, G., Roberts, E., Springell, K., Karbani, G., Jafri, H., Mannan, J., Raashid, Y. and Al-Gazali, L., 'An SCN9A channelopathy causes congenital inability to experience pain', *Nature*, 444, 2006, pp.894–8

[19] Andresen, T., Lunden, D., Drewes, A. M. and Arendt-Nielsen, L., 'Pain sensitivity and experimentally induced sensitisation in red haired females', *Scandinavian Journal of Pain*, 2(1), 2011, pp.3–6

[20] 'Paget, Henry William, first Marquis of Anglesey (1768–1854)', *Oxford Dictionary of National Biography*, Oxford University Press, 2004 (online edition)

[21] Titus Lucretius Carus, *Lucretius: The Nature of Things,* trans. Stallings, A. E., Penguin Classics, 2007

[22] Denk, F., Crow, M., Didangelos, A., Lopes, D. M. and McMahon, S. B., 'Persistent alterations in microglial enhancers in a model of chronic pain', *Cell Reports*, 15(8), 2016, pp.1771–81

[23] de Montaigne, Michel, *The Complete Essays*, trans. Screech, M. A., Penguin Classics, 1993, Book 3, Chapter 13

[24] Handwerker, H. O., Magerl, W., Klemm, F., Lang, E. and Westerman, R. A., 'Quantitative evaluation of itch sensation', *Fine Afferent Nerve Fibers and Pain,* eds. Schmidt, R.F., Schaible, H.-G., Vahle-Hinz, C., VCH Verlagsgesellschaft, Weinheim, 1987, pp.462–73

[25] Pitake, S., DeBrecht, J. and Mishra, S. K., 'Brain natriuretic peptide-expressing sensory neurons are not involved in acute, inflammatory, or neuropathic pain', *Molecular Pain*, 13, 2017

[26] Holle, H., Warne, K., Seth, A. K., Critchley, H. D. and Ward, J., 'Neural basis of contagious itch and why some people are more prone to it', *Proceedings of the National Academy of Sciences*, 109(48), 2012, pp.19816–21

[27] Lloyd, D. M., Hall, E., Hall, S. and McGlone, F. P., 'Can itch-related visual stimuli alone provoke a scratch response in healthy individuals?', *British Journal of Dermatology*, 168(1), 2013, pp.106–11

[28] Yu, Y. Q., Barry, D. M., Hao, Y., Liu, X. T. and Chen, Z. F., 'Molecular and neural basis of contagious itch behavior in mice', *Science*, 355(6329), 2017, pp.1072–6

[29] Jourard, S. M., 'An exploratory study of body-accessibility', *British Journal of Clinical Psychology*, 5(3), 1966, pp.221–31

[30] Ackerman, J. M., Nocera, C. C. and Bargh, J. A., 'Incidental haptic sensations influence social judgments and decisions', *Science*, 328(5986), 2010, pp.1712–5

[31] Levav, J. and Argo, J. J., 'Physical contact and financial risk taking', *Psychological Science*, 21(6), 2010, pp.804–10

[32] Ackerman, J. M., Nocera, C. C. and Bargh, J. A., 'Incidental haptic sensations influence social judgments and decisions', *Science*, 328(5986), 2010, pp.1712–15

[33] Kraus, M. W., Huang, C. and Keltner, D., 'Tactile communication, cooperation, and performance: an ethological study of the NBA', *Emotion*, 10(5), 2010, p.745

[34] Hertenstein, M. J., Holmes, R., McCullough, M. and Keltner, D., 'The communication of emotion via touch', *Emotion*, 9(4), 2009, p.566

[35] Brentano, R. 'Reviewed Work: *The Chronicle of Salimbene de Adam* by Salimbene de Adam, Joseph L. Baird, Giuseppe Baglivi, John Robert Kane', *The Catholic Historical Review*, 74(3), 1988, pp.466–7

[36] Field, T. M., *Touch in Early Development*, Psychology Press, 2014

[37] Pollak, S. D., Nelson, C. A., Schlaak, M. F., Roeber, B. J., Wewerka, S. S., Wiik, K. L., Frenn, K. A., Loman, M. M. and Gunnar, M. R., 'Neurodevelopmental effects of early deprivation in postinstitutionalized children', *Child Development*, 81(1), 2010, pp.224–36

[38] Rey Sanabria, E. and Gómez, H. M., 'Manejo Racional del Niño Prematuro [Rational management of the premature child]', *Fundación Vivir*, Bogotá, Colombia, 1983, pp.137–51

[39] Lawn, J. E., Mwansa-Kambafwile, J., Horta, B. L., Barros, F. C. and Cousens, S., ' "Kangaroo mother care" to prevent neonatal deaths due to preterm birth complications', *International Journal of Epidemiology*, 39 (Supplement 1), 2010, pp.i144–54

[40] Charpak, N., Tessier, R., Ruiz, J. G., Hernandez, J. T., Uriza, F., Villegas, J., Nadeau, L., Mercier, C., Maheu, F., Marin, J. and Cortes, D., 'Twenty-year follow-up of kangaroo mother care versus traditional care', *Pediatrics*, 2016, p.e20162063

[41] Sloan, N. L., Ahmed, S., Mitra, S. N., Choudhury, N., Chowdhury, M., Rob, U. and Winikoff, B., 'Community-based kangaroo mother care to prevent neonatal and infant mortality: a randomized, controlled cluster trial', *Pediatrics*, 121(5), 2008, pp.e1047–59

[42] Coan, J. A., Schaefer, H. S. and Davidson, R. J., 'Lending a hand: social regulation of the neural response to threat', *Psychological Science*, 17(12), 2006, pp.1032–9

[43] Holt-Lunstad, J., Birmingham, W. A. and Light, K. C., 'Influence of a "warm touch" support enhancement intervention among married couples

on ambulatory blood pressure, oxytocin, alpha amylase, and cortisol', *Psychosomatic Medicine*, 70(9), 2008, pp.976–85

[44] Field, T. M., 'Massage therapy research review', *Complementary Therapies in Clinical Practice*, 20(4), 2014, pp.224–9

[45] Kim, H. K., Lee, S. and Yun, K. S., 'Capacitive tactile sensor array for touch screen application', *Sensors and Actuators A: Physical*, 165(1), 2011, pp.2–7

[46] Jiménez, J., Olea, J., Torres, J., Alonso, I., Harder, D. and Fischer, K., 'Biography of Louis Braille and invention of the Braille alphabet', *Survey of Ophthalmology*, 54(1), 2009, pp.142–9

[47] Choi, S. and Kuchenbecker, K. J., 'Vibrotactile display: Perception, technology, and applications', *Proceedings of the IEEE*, 101(9), 2013, pp.2093–104

[48] Culbertson, H. and Kuchenbecker, K. J., 'Importance of Matching Physical Friction, Hardness, and Texture in Creating Realistic Haptic Virtual Surfaces', *IEEE Transactions on Haptics*, 10(1), 2017, pp.63–74

[49] Saal, H. P., Delhaye, B. P., Rayhaun, B. C. and Bensmaia, S. J., 'Simulating tactile signals from the whole hand with millisecond precision', *Proceedings of the National Academy of Sciences*, 114(28), 2017, pp.E5693–E5702

[50] Wu, W., Wen, X. and Wang, Z. L., 'Taxel-addressable matrix of vertical-nanowire piezotronic transistors for active and adaptive tactile imaging', *Science*, 340(6135), 2013, pp.952–7

[51] Yin, J., Santos, V. J. and Posner, J. D., 'Bioinspired flexible microfluidic shear force sensor skin', *Sensors and Actuators A: Physical*, 264, 2017, pp.289–97

7 Psychological Skin

[1] Koblenzer, C. S., 'Dermatitis artefacta: clinical features and approaches to treatment', *American Journal of Clinical Dermatology*, 1(1), 2000, pp.47–55

[2] Deweerdt, S., 'Psychodermatology: an emotional response', *Nature*, 492(7429), 2012, pp.S62–3

[3] Evers, A. W. M., Verhoeven, E. W. M., Kraaimaat, F. W., De Jong, E. M. G. J., De Brouwer, S. J. M., Schalkwijk, J., Sweep, F. C. G. J. and Van De Kerkhof, P. C. M., 'How stress gets under the skin: cortisol and stress reactivity in psoriasis', *British Journal of Dermatology*, 163(5), 2010, pp.986–91

[4] Pavlovic, S., Daniltchenko, M., Tobin, D. J., Hagen, E., Hunt, S. P., Klapp, B. F., Arck, P. C. and Peters, E. M., 'Further exploring the brain–skin connection: stress worsens dermatitis via substance P-dependent neurogenic inflammation in mice', *Journal of Investigative Dermatology*, 128(2), 2008, pp.434–46

[5] Peters, E. M., 'Stressed skin? – a molecular psychosomatic update on stress – causes and effects in dermatologic diseases', *Journal der Deutschen Dermatologischen Gesellschaft*, 14(3), 2016, pp.233–52

[6] Naik, S., Larsen, S. B., Gomez, N. C., Alaverdyan, K., Sendoel, A., Yuan, S., Polak, L., Kulukian, A., Chai, S. and Fuchs, E., 'Inflammatory memory sensitizes skin epithelial stem cells to tissue damage', *Nature*, 550(7677), 2017, p.475

[7] Felice, C., *Here Are the Young Men* (photography series), 2009–2010

[8] Schwartz, J., Evers, A. W., Bundy, C. and Kimball, A. B., 'Getting under the skin: report from the International Psoriasis Council Workshop on the role of stress in psoriasis', *Frontiers in Psychology*, 7, 2016, p.87

[9] Bewley, A. P., 'Snapshot survey of dermatologists' reports of skin disease following the financial crisis of 2007–2008', *British Skin Foundation*, 2012

[10] Dhabhar, F. S., 'Acute stress enhances while chronic stress suppresses skin immunity: the role of stress hormones and leukocyte trafficking', *Annals of the New York Academy of Sciences*, 917(1), 2000, pp.876–93

[11] Kabat-Zinn, J., Wheeler, E., Light, T., Skillings, A., Scharf, M. J., Cropley, T. G., Hosmer, D. and Bernhard, J. D., 'Influence of a mindfulness meditation-based stress reduction intervention on rates of skin clearing in patients with moderate to severe psoriasis undergoing photo therapy (UVB) and photochemotherapy (PUVA)', *Psychosomatic Medicine*, 60(5), 1998, pp.625–32

[12] Dijk, C., Voncken, M. J. and de Jong, P. J., 'I blush, therefore I will be judged negatively: influence of false blush feedback on anticipated others' judgments and facial coloration in high and low blushing-fearfuls', *Behaviour Research and Therapy*, 47(7), 2009, pp.541–7

[13] Dijk, C., de Jong, P. J. and Peters, M. L., 'The remedial value of blushing in the context of transgressions and mishaps', *Emotion*, 9(2), 2009, p.287

[14] Dijk, C. and de Jong, P. J., 'Blushing-fearful individuals overestimate the costs and probability of their blushing', *Behaviour Research and Therapy*, 50(2), 2012, pp.158–62

[15] Mirick, D. K., Davis, S. and Thomas, D. B., 'Antiperspirant use and the risk of breast cancer', *Journal of the National Cancer Institute*, 94(20), 2002, pp.1578–80

[16] Willhite, C. C., Karyakina, N. A., Yokel, R. A., Yenugadhati, N., Wisniewski, T. M., Arnold, I. M., Momoli, F. and Krewski, D., 'Systematic review of potential health risks posed by pharmaceutical, occupational and consumer exposures to metallic and nanoscale aluminum, aluminum oxides, aluminum hydroxide and its soluble salts', *Critical Reviews in Toxicology*, 44(sup4), 2014, pp.1–80

[17] Hermann, L. and Luchsinger, B., 'Über die Secretionsströme der Haut bei der Katze [On the sweat currents on the skin of cats]', *Pflügers Archiv European Journal of Physiology*, 17(1), 1878, pp.310–19

[18] *Idaho State Journal*, 9 November 1977, p.32

[19] Larson, J. A., Haney, G. W. and Keeler, L., *Lying and its detection: A study of deception and deception tests*, University of Chicago Press, 1932, p.99

[20] Inbau, F. E., 'Detection of deception technique admitted as evidence', *Journal of Criminal Law and Criminology (1931-51)*, 26(2), 1935, pp.262–70

[21] Santos, F., 'DNA evidence frees a man imprisoned for half his life', *New York Times*, 1 September 2006

[22] Goldstein, A., 'Thrills in response to music and other stimuli', *Physiological Psychology*, 8(1), 1980, pp.126–9

[23] Timmers, R., and Loui, P., 'Music and Emotion', *Foundations in Music Psychology*, eds. Rentfrow, P. J, and Levitin, D. J., MIT Press, 2019, pp.783–826

[24] Blood, A. J. and Zatorre, R. J., 'Intensely pleasurable responses to music correlate with activity in brain regions implicated in reward and emotion', *Proceedings of the National Academy of Sciences*, 98(20), 2001, pp.11818–23

[25] Hongbo, Y., Thomas, C. L., Harrison, M. A., Salek, M. S. and Finlay, A. Y., 'Translating the science of quality of life into practice: what do dermatology life quality index scores mean?', *Journal of Investigative Dermatology*, 125(4), 2005, pp.659–64

[26] Ramrakha, S., Fergusson, D. M., Horwood, L. J., Dalgard, F., Ambler, A., Kokaua, J., Milne, B. J. and Poulton, R., 'Cumulative mental health consequences of acne: 23-year follow-up in a general population birth cohort study', *The British Journal of Dermatology*, 2015

[27] British Skin Foundation Teenage Acne Survey 2014–2017 press release, '3 in 5 teenagers say acne affects self confidence', 2017

[28] Chiu, A., Chon, S. Y. and Kimball, A. B., 'The response of skin disease to stress: changes in the severity of acne vulgaris as affected by examination stress', *Archives of Dermatology*, 139(7), 2003, pp.897–900

[29] Böhm, D., Schwanitz, P., Stock Gissendanner, S., Schmid-Ott, G. and Schulz, W., 'Symptom severity and psychological sequelae in rosacea: results of a survey', *Psychology, Health & Medicine*, 19(5), 2014, pp.586–91

[30] Sharma, N., Koranne, R. V. and Singh, R. K., 'Psychiatric morbidity in psoriasis and vitiligo: a comparative study', *The Journal of Dermatology*, 28(8), 2001, pp.419–23

[31] Tsakiris, M. and Haggard, P., 'The rubber hand illusion revisited: visuotactile integration and self-attribution', *Journal of Experimental Psychology: Human Perception and Performance*, 31(1), 2005, p.80

[32] Lovato, L., Ferrão, Y. A., Stein, D. J., Shavitt, R. G., Fontenelle, L. F., Vivan, A., Miguel, E. C. and Cordioli, A. V., 'Skin picking and trichotillomania in adults with obsessive-compulsive disorder', *Comprehensive Psychiatry*, 53(5), 2012, pp.562–68

[33] Bjornsson, A. S., Didie, E. R. and Phillips, K. A., 'Body dysmorphic disorder', *Dialogues in Clinical Neuroscience*, 12(2), 2010, p.221

[34] Kim, D. I., Garrison, R. C. and Thompson, G., 'A near fatal case of pathological skin picking', *The American Journal of Case Reports*, 14, 2013, pp.284–7

8 Social Skin

[1] Orange, C., *The Treaty of Waitangi*, Bridget Williams Books, 2015

[2] Cook, J., *Captain Cook's Journal During His First Voyage Round the World, Made in HM* Bark Endeavour, *1768–71*, Cambridge University Press, 2014

[3] News stories from the universities of Birmingham and Oxford regarding the return of Maori heads: www.birmingham.ac.uk/news/latest/2013/10/

Maori-remains-make-the-long-journey-to-their-ancestral-home; www.glam.ox.ac.uk/article/repatriation-maori-ancestral-remains

[4] Samuel O'Reilly's patent for a tattoo machine: *S. F. O'Reilly, Tattooing Machine, No. 464,801, Patented Dec. 8, 1891*

[5] Othman, J., Robbins, E., Lau, E. M., Mak, C. and Bryant, C., 'Tattoo pigment-induced granulomatous lymphadenopathy mimicking lymphoma', *Annals of Internal Medicine*, 2017

[6] Huq, R., Samuel, E. L., Sikkema, W. K., Nilewski, L. G., Lee, T., Tanner, M. R., Khan, F. S., Porter, P. C., Tajhya, R. B., Patel, R. S. and Inoue, T., 'Preferential uptake of antioxidant carbon nanoparticles by T lymphocytes for immunomodulation', *Scientific Reports*, 6, 2016, article 33808

[7] Brady, B. G., Gold, H., Leger, E. A. and Leger, M. C., 'Self-reported adverse tattoo reactions: a New York City Central Park study', *Contact Dermatitis*, 73(2), 2015, pp.91–9

[8] Kreidstein, M. L., Giguere, D. and Freiberg, A., 'MRI interaction with tattoo pigments: case report, pathophysiology, and management', *Plastic and Reconstructive Surgery*, 99(6), 1997, pp.1717–20

[9] Schreiver, I., Hesse, B., Seim, C., Castillo-Michel, H., Villanova, J., Laux, P., Dreiack, N., Penning, R., Tucoulou, R., Cotte, M. and Luch, A., 'Synchrotron-based ν-XRF mapping and μ-FTIR microscopy enable to look into the fate and effects of tattoo pigments in human skin', *Scientific Reports*, 7(1), 2017, article 11395

[10] Laux, P., Tralau, T., Tentschert, J., Blume, A., Al Dahouk, S., Bäumler, W., Bernstein, E., Bocca, B., Alimonti, A., Colebrook, H. and de Cuyper, C., 'A medical-toxicological view of tattooing', *The Lancet*, 387(10016), 2016, pp.95–402

[11] Brady, B. G., Gold, H., Leger, E. A. and Leger, M. C., 'Self-reported adverse tattoo reactions: a New York City Central Park study', *Contact Dermatitis*, 73(2), 2015, pp.91–9

[12] Liszewski, W., Kream, E., Helland, S., Cavigli, A., Lavin, B. C. and Murina, A., 'The demographics and rates of tattoo complications, regret, and unsafe tattooing practices: a cross-sectional study', *Dermatologic Surgery*, 41(11), 2015, pp.1283–89

[13] Ephemeral Tattoos: www.ephemeraltattoos.com

[14] Kim, J., Jeerapan, I., Imani, S., Cho, T. N., Bandodkar, A., Cinti, S., Mercier, P. P. and Wang, J., 'Noninvasive alcohol monitoring using a wearable tattoo-based iontophoretic-biosensing system', *ACS Sensors*, 1(8), 2016, pp.1011–19

[15] Bareket, L., Inzelberg, L., Rand, D., David-Pur, M., Rabinovich, D., Brandes, B. and Hanein, Y., 'Temporary-tattoo for long-term high fidelity biopotential recordings', *Scientific Reports*, 6, 2016, article 25727

[16] Garcia, S. O., Ulyanova, Y. V., Figueroa-Teran, R., Bhatt, K. H., Singhal, S. and Atanassov, P., 'Wearable sensor system powered by a biofuel cell for detection of lactate levels in sweat', *ECS Journal of Solid State Science and Technology*, 5(8), 2016, pp.M3075–81

[17] Liu, X., Yuk, H., Lin, S., Parada, G. A., Tang, T. C., Tham, E., de la Fuente-Nunez, C., Lu, T. K. and Zhao, X., '3D printing of living responsive materials and devices', *Advanced Materials*, 30(4), 2018

[18] Samadelli, M., Melis, M., Miccoli, M., Vigl, E. E. and Zink, A. R., 'Complete mapping of the tattoos of the 5300-year-old Tyrolean Iceman', *Journal of Cultural Heritage*, 16(5), 2015, pp.753–8

[19] Krutak, L. F., *Spiritual Skin – Magical Tattoos and Scarification: Wisdom. Healing. Shamanic power. Protection*. Edition Reuss, 2012

[20] Krutak, L., 'The cultural heritage of tattooing: a brief history', *Tattooed Skin and Health*, 48, 2015, pp.1–5

[21] Krutak, L., 'The cultural heritage of tattooing: a brief history', *Tattooed Skin and Health*, (48), 2015, pp.1–5

[22] Lynn, C. D., Dominguez, J. T. and DeCaro, J. A., 'Tattooing to "toughen up": tattoo experience and secretory immunoglobulin A', *American Journal of Human Biology*, 28(5), 2016, pp.603–9

[23] Chiu, Y. N., Sampson, J. M., Jiang, X., Zolla-Pazner, S. B. and Kong, X. P., 'Skin tattooing as a novel approach for DNA vaccine delivery', *Journal of Visualized Experiments*, 68, 2012

[24] Landeg, S. J., Kirby, A. M., Lee, S. F., Bartlett, F., Titmarsh, K., Donovan, E., Griffin, C. L., Gothard, L., Locke, I. and McNair, H. A., 'A randomized control trial evaluating fluorescent ink versus dark ink tattoos for breast radiotherapy', *British Journal of Radiology*, 89(1068), 2016, p.20160288

[25] Wolf, E. K. and Laumann, A. E., 'The use of blood-type tattoos during the Cold War', *Journal of the American Academy of Dermatology*, 58(3), 2008, pp.472–6

[26] Holt, G. E., Sarmento, B., Kett, D. and Goodman, K. W., 'An unconscious patient with a DNR tattoo', *New England Journal of Medicine*, 377(22), 2017, pp.2192–3

[27] Banks, J., *Journal of the Right Hon. Sir Joseph Banks: During Captain Cook's First Voyage in H.M.S.* Endeavour *in 1768–71*, Cambridge University Press, 2011

9 The Skin That Separates

[1] International Federation of Red Cross and Red Crescent Societies, 'Through albino eyes: the plight of albino people in Africa's Great Lakes region and a Red Cross response', 2009

[2] Jablonski, N. G. and Chaplin, G., 'Human skin pigmentation as an adaptation to UV radiation', *Proceedings of the National Academy of Sciences*, 107 (Supplement 2), 2010, pp.8962–8

[3] Bauman, Z., 'Modernity and ambivalence', *Theory, Culture & Society*, 7(2–3), 1990. pp.143–69

[4] Yudell, M., Roberts, D., DeSalle, R. and Tishkoff, S., 'Taking race out of human genetics', *Science*, 351(6273), 2016, pp.564–5

[5] Crawford, N. G., Kelly, D. E., Hansen, M. E., Beltrame, M. H., Fan, S., Bowman, S. L., Jewett, E., Ranciaro, A., Thompson, S., Lo, Y. and Pfeifer, S. P., 'Loci associated with skin pigmentation identified in African populations', *Science*, 358(6365), 2017. p.eaan8433

[6] Roncalli, R. A., 'The history of scabies in veterinary and human medicine from biblical to modern times', *Veterinary Parasitology*, 25(2), 1987, pp.193–8

[7] Jenner, E., *An Inquiry into The Causes and Effects of the Variolae Vaccinae, A Disease Discovered in Some of the Western Counties Of England, Particularly Gloucestershire, and Known By The Name of The Cow Pox*, 1800

[8] Riedel, S., 'Edward Jenner and the history of smallpox and vaccination', *Baylor University Medical Center Proceedings*, 18(1), 2005, p.21

[9] Kricker, A., Armstrong, B. K., English, D. R. and Heenan, P. J., 'A dose-response curve for sun exposure and basal cell carcinoma', *International Journal of Cancer*, 60(4), 1995, pp.482–8

[10] Loewenthal, L. J. A., 'Daniel Turner and "De Morbis Cutaneis"', *Archives of Dermatology*, 85(4), 1962, pp.517–23

[11] Flotte, T. J. and Bell, D. A., 'Role of skin lesions in the Salem witchcraft trials', *The American Journal of Dermatopathology*, 11(6), 1989, pp.582–7

[12] Karen Hearn, 'Why do so many people want their moles removed?', BBC News, 11 November 2015

[13] King, D. F. and Rabson, S. M., 'The discovery of Mycobacterium leprae: A medical achievement in the light of evolving scientific methods', *The American Journal of Dermatopathology*, 6(4), 1984, pp.337–44

[14] Monot, M., Honoré, N., Garnier, T., Araoz, R., Coppée, J. Y., Lacroix, C., Sow, S., Spencer, J. S., Truman, R. W., Williams, D. L. and Gelber, R., 'On the origin of leprosy', *Science*, 308(5724), 2005, pp.104–42

[15] Fine, P. E., Sterne, J. A., Pönnighaus, J. M. and Rees, R. J., 'Delayed-type hypersensitivity, mycobacterial vaccines and protective immunity', *The Lancet*, 344(8932), 1994, pp.1245–9

[16] Doniger, W., *The Laws of Manu*, Penguin, 1991

[17] Wright, H. P., *Leprosy – An Imperial Danger*, Churchill, 1889

[18] Herman, R. D. K., 'Out of sight, out of mind, out of power: leprosy, race and colonization in Hawaii', *Journal of Historical Geography*, 27(3), 2001, pp.319–37

[19] Horman, W. *Vulgaria Puerorum*, 1519

[20] Blomfield, A., 'Rwandan police crack down on harmful skin bleaching products', *Daily Telegraph*, 10 January 2019

10 Spiritual Skin

[1] Connor, S., *The Book of Skin*, Cornell University Press, 2004

[2] McNeley, J. K., *Holy Wind in Navajo Philosophy*, University of Arizona Press, 1981

[3] Tweed, T. A., *Crossing and Dwelling: A Theory of Religion*, Harvard University Press, 2009

[4] Allen, P. L., *The Wages of Sin: Sex and Disease, Past and Present*, University of Chicago Press, 2000

[5] Piper, J., *Stripped in Shame, Clothed in Grace*, 2007

[6] Lynch, P. A. and Roberts, J., *Native American Mythology A to Z*, Infobase Publishing, 2004

[7] Benthien, C., *Skin: On the Cultural Border Between Self and the World*, Columbia University Press, 2002

[8] Anzieu, D., *The Skin-Ego*, Karnac Books, 2016

[9] Bachelard, G., *The Poetics of Space*, vol. 330, Beacon Press, 1994

[10] Foucault, M., 'Technologies of the Self', *Technologies of the Self: A seminar with Michel Foucault*, University of Massachusetts Press, 1988, pp.16–49

[11] Dudley-Edwards, O., *Burke and Hare*, Birlinn, 2014

[12] Bailey, B., *Burke and Hare: The Year of the Ghouls*, Mainstream, 2002

Acknowledgements

This book is dedicated to the millions across the world who suffer in, or for, their skin. Some of these people kindly told me their stories, teaching me something of the despairs and the joys of humanity. Without them this book would be no more than a flimsy health-information leaflet.

I have wanted to write a book about science and medicine since I was very young, and I want to thank everyone who has made this possible. I'm indebted to my indefatigable editor at Transworld, Andrea Henry, and George Gibson at Grove Atlantic, for their guidance and hard work. Thanks also to the magnificent team at Transworld: Phil Lord, Tom Hill, Kate Samano, Richard Shailer, Alex Newby and Doug Young.

To my excellent agent Charlie Viney, for his help, wisdom and faith in the proposal from the very beginning.

This book would never have got off the ground without the generous travel and research funding from some wonderful charities and organizations. I'd like to thank the British Association of Dermatologists, the Thesiger Award committee, Richard Sykes and the Waltham St Lawrence Charities, the St Francis Leprosy Guild, EMMS International and the Kohima Educational Trust.

To the dermatologists and doctors in Birmingham, Oxford and London who patiently taught, advised and inspired me: Alexa Shipman, Sajjad Rajpar, James Halpern, Ser-Ling Chua, Tom Tull, Mary Glover, Chris Bunker, Terence Ryan, the DermSoc and DermSchool committees at the British Association of Dermatologists and others across the world, particularly those at the Regional Dermatology Training Centre in Tanzania and the Naga Hospital in India.

To the cutaneous immunology group in Oxford: particularly Graham Ogg for his generous and encouraging tutorship, and Clare Hardman and Janina Nahler for putting up with my precarious pipetting.

To the DermSoc team at the British Association of Dermatologists for being the first to open my eyes to the little-known wonders of the skin: Siu Tsang, Ketaki Bhate, Bernard Ho, Katie Farquhar, Anna Ascott, Natasha Lee and Sophia Haywood.

To Colin Thubron and Margreta de Grazia, for reading a couple of thousand words of my writing a few years ago and saying 'there's a book in this'.

To other supporters and role models: John Beale, Jamie Mills, George Fussey, Glynn Harrison and Kate Thomas.

To Hannah. I'm sorry I mentioned the book the first time we met and have droned on about it ever since. I don't know what I would have done without your editor's eye and your patient heart.

To my unofficial (but no less ruthless) editors: my family. To Rob for being not only a great support but also my greatest writing role model, to Hannah for invaluable critique and advice (and teaching me to say 'onchocerciasis') and to Phin for encouraging me, against all of your brotherly instincts.

And finally to the scientists, writers and historians acknowledged in the bibliography. These references are a small snapshot of the collective work of researchers across the globe who dedicate their lives to the advancement of human knowledge and the pursuit of truth. These are the giants on whose shoulders I stand.

Index

Figures in **bold** refer to pages with illustrations.